Telephone Renewals: 0117 32 82092 (24 hours)
Library Web Address: www.uwe.ac.uk/library

D1355732

Carbon Reduction:
Policies, Strategies, and Technologies

Carbon Reduction:
Policies, Strategies, and Technologies

Stephen A. Roosa, Ph.D.

Arun G. Jhaveri, Ed.D.

THE FAIRMONT PRESS, INC.

CRC Press
Taylor & Francis Group

Library of Congress Cataloging-in-Publication Data

Roosa, Stephen A., 1952-
    Carbon reduction : policies, strategies, and technologies / Stephen A. Roosa, Arun G. Jhaveri.
       p. cm.
    Includes bibliographical references and index.
    ISBN-10: 0-88173-603-1 (alk. paper)
    ISBN-10: 0-88173-604-X (electronic)
    ISBN-13: 978-1-4200-8382-8 (Taylor & Francis : alk. paper)
     1. Carbon dioxide mitigation. I. Jhaveri, Arun G., 1938- II. Title.

TD885.5.C3R66 2009
363.738'746--dc22

2009015128

Published by The Fairmont Press, Inc.
700 Indian Trail
Lilburn, GA 30047
tel: 770-925-9388; fax: 770-381-9865
http://www.fairmontpress.com

Distributed by Taylor & Francis Ltd.
6000 Broken Sound Parkway NW, Suite 300
Boca Raton, FL 33487, USA
E-mail: orders@crcpress.com

Distributed by Taylor & Francis Ltd.
23-25 Blades Court
Deodar Road
London SW15 2NU, UK
E-mail: uk.tandf@thomsonpublishingservices.co.uk

Printed in the United States of America
10 9 8 7 6 5 4 3 2 1

10: 0-88173-603-1 (The Fairmont Press, Inc.)
13: 978-1-4200-8382-8 (Taylor & Francis Ltd.)

While every effort is made to provide dependable information, the publisher, authors, and editors cannot be held responsible for any errors or omissions.

Table of Contents

Preface

The Association of Energy Engineers (AEE), a nonprofit professional society of 8,500 members, presented the first World Energy Engineering Congress in 1978. At that meeting, former astronaut Alfred Worden presented the keynote address and stated:

> "The human appetite for energy appears to be insatiable. There are good reasons for the continued growth of energy consumption in the future: the survival needs of the underprivileged millions, increased adult life expectations stemming from the population boom... and the desirable goal of improved quality of life for everyone. But there are equally good reasons for carefully examining what the consequences of energy growth will be after present consumption rates have quadrupled. When do we melt the polar icecaps? How much land can we afford to set aside for energy plants? How much photosynthetic smog can we tolerate?"

For more than thirty-one years the Association of Energy Engineers (AEE) has played a major role in developing the "Green Industry" and encouraging the adoption of energy efficient and renewable energy technologies for sustainable development. Finally, here is an indispensable reference book to guide the reader through the maze of issues associated with approaches to carbon reduction. We are pleased that two key AEE members, Dr. Stephen Roosa and Dr. Arun Jhaveri have authored this groundbreaking book, *Carbon Reduction: Policies, Strategies and Technologies*.

The new corporate ethic is to reduce our carbon footprint. This book will play a major role in achieving this goal.

Albert Thumann, P.E, CEM
Executive Director
Association of Energy Engineers

Acknowledgements

In just the last decade, the need to reduce carbon emissions has become one of the most pressing environmental concerns. In our investigations of this topic, we have had the opportunity to learn and share experiences with people who have helped us understand the importance of this seminal issue. Our work has been dedicated to addressing the linkage between energy and carbon reduction while exploring the policies and technologies that offer hope that a solution can be found in our lifetimes. Many people and organizations, from a wide range of disciplines and from all over the world, have aided our efforts. We are humbled by their support, research and dedication to this topic.

Researching carbon reduction—truly a fast-moving international phenomenon—quickly became a travel-intensive endeavor. We would like to thank Al Thumann, Executive Director of the Association of Energy Engineers (AEE) who contributed the Preface to this book and who along with his outstanding staff organized topical AEE conferences with tracks and seminars that we attended in Seattle, Austin, Washington, DC, Atlantic City and Atlanta. It was at these presentations that we learned from various experts in the field how carbon reduction technologies were being developed and deployed. We thank Dr. Albin Zsebig, from the Budapest University of Science and Technology, who invited Dr. Roosa to Hungary on several occasions to co-chair international conferences that he had organized, allowing an understanding of how Europeans were approaching climate change and carbon reduction issues. Also, Dr. John Gilderbloom, who invited Dr. Roosa to co-chair conferences in Honolulu and Havana, and to present his research at a conference in Amsterdam in 2008, which provided opportunities to study carbon reduction solutions in Holland. We thank him for his encouragement and friendship. Dr. Roosa also thanks April Li and other members of the organizing committee of the Electrical and Mechanical Safety and Energy Efficiency Symposium 2007 in Hong Kong, who had invited him as keynote speaker, which provided an opportunity to research carbon reduction approaches there.

Without the invaluable efforts of the book's contributors, the topic of carbon reduction could not have been so rigorously addressed. Our sincere thanks go to:

- Danielle E. Miller, who worked tirelessly over the course of an extended summer, offering invaluable assistance with topical research, preliminary editing and the updating of chapter drafts; her efforts combined with her insights helped make this book all that it is while somehow managing to keep us on task and on schedule. She also helped draft the introduction, and chapter summaries, compiled references, and contributed sections to Chapter 8. This vibrant lady is truly the heart and spirit of our book. We also thank Vanderbilt University for arranging funding for her internship with us.

- Matt Hanka, a former student of Dr. Roosa at the University of Louisville who contributed to Chapter 3 and provided perspective and valuable insights regarding the discussions of the Kyoto and Montreal Protocols.

Dr. Roosa also thanks his many former graduate and undergraduate environmental policy students at the University of Louisville. It was in those courses that fruitful discussions regarding carbon reduction first occurred, forming the ideas that first generated Dr. Roosa's interest this topic.

Dr. Arun Jhaveri's inspirations in the fields of energy, environment, sustainability, climate change, and technology, have been derived from a variety of personal and professional work sources:

- Creation of the U.S. government's new Environmental Protection Agency (USEPA) and the National Environmental Policy Act (NEPA) in the 1970s;
- Being the first mayor of the new sustainable City of Burien in Washington State;
- Representing both the local and federal governments as U.S. delegate to the 1997 Kyoto Protocol deliberations in Japan and to the 2002 United Nation's World Summit on Sustainable Development (WSSD) in Johannesburg, South Africa; and
- Completing his Doctoral Dissertation entitled "Effective Leadership for Sustainable Development in the Public Sector" at Seattle University in 2006;

- Working for the U.S. federal government for 24 years, including 8 years with the U.S. Army Corps of Engineers as engineering project manager, and 16 years with the U.S. Department of Energy as regional manager.

These experiences and expertise were complemented by Dr. Jhaveri's active involvement in national and international professional and technical organizations including the Association of Energy Engineers (AEE), National League of Cities (NLC), and the International Leadership Association (ILA). Dr. Jhaveri received significant moral support from many colleagues, friends, and family members on this book.

It was only fitting that Bill Payne, the editor of Dr. Roosa's *Sustainable Development Handbook*, should be asked to work on this project. We thank him for his outstanding work and contributions to this book.

We have appreciated and been motivated by the support of our friends, family, and professional counterparts that have assisted us in so many ways. We thank them for their support.

Introduction

Make no mistake—we have now reached a crossroad in human history. We have finally determined that mankind has the power not only to change the world's environment, but also the ability do so in a manner that is so ecologically disruptive as to cause damage that threatens our very survival. The new overarching environmental problem is the rapid increase in atmospheric carbon levels. A policy of ignoring the problem of atmospheric carbon concentrations places us at ever-greater risk, creates unmanageable uncertainties, and will force us to make unforeseen sacrifices in the future. We are entering a time when our options to react are diminished by the infrastructure we create. Without concerted action on a global scale, reducing the impact of this threat will be increasingly difficult. To face this crisis, we find that we must revamp our governmental policies and institute a new range of programs. Simultaneously, we must develop and implement new energy technologies that enable permanent carbon reductions. The complexity of this problem is confounded by the world's developmental needs, growing populations, progressive social goals, market uncertainties, resource constraints and untested technologies. The importance of addressing increasing atmospheric carbon emissions cannot be understated—both the sustainability of the planet's ecosystems and the survivability of mankind are at stake.

A new era is upon us—one that will focus on energy technologies and how we transition our economies to reduce carbon emissions by both reducing the combustion of carbon and preventing its release into the Earth's atmosphere. *Carbon Reduction: Policies, Strategies and Technologies* provides an overview of carbon reduction approaches as a means of addressing one of the world's most important sustainability issues. The purpose of this book is to address the complexity of these problems and offer workable solutions. Eventually, individual social responsibility must become the cornerstone of these solutions.

This book summarizes the history, policies and agenda that are the key building blocks in the implementation of carbon reduction programs. The policies that are adopted provide a basis for the development of programs that can be implemented on many levels. Workable technologies are deployed that have potential to achieve carbon reduction goals. This book also considers management approaches and the means of financing carbon reduction projects. Technologies are discussed that have potential to achieve carbon reduction goals—if we choose to implement them quickly and on an appropriate scale.

Chapter 1 verifies the significance of increasing levels of atmospheric CO_2 and their relationship to sustainable development. The anticipated effects of CO_2 and greenhouse gas (GHG) emissions are illustrated, as inferred through broad scientific consensus, as far-reaching, negative influences on future generations. Information provided in "The Importance of Reducing Carbon Emissions" attributes excessive greenhouse gas and CO_2 emissions to the evolution of human societies. Underlying causes and trends are identified and explained in reference to the economic, political, and social context. These developments emphasize the world's environmental instability and difficulties in resolving issues associated with climate change.

Excessive generation of carbon emissions is directly linked to global climate change. The fragility of the earth's ecosystems and the potential irreversibility of climate change are identified as a global environmental threat in this chapter, calling for concerted action on every possible scale. New strategies, programs and technologies are presented for their potential to mitigate carbon emissions. Most significantly, this chapter demonstrates that synchronized, comprehensive efforts are needed.

Chapter 2 begins by implicating the connection between CO_2 levels in the atmosphere and the average Earth temperature. It provides a relevant history of rising anthropogenic CO_2 emissions and correlating temperatures. The threshold for CO_2 levels is identified—beyond which environmental damage will be irreparable—as well as efforts to forestall its occurrence. Related technologies and strategies for "low carbon solutions" are also identified in this chapter.

The problematic nature of CO_2 is explained along with a description of its various impacts. In terms of biology, the analogy of Greenhouse Earth is discussed, and the ability of CO_2 to trap additional heat in the atmosphere. Another issue is the disproportionate generation of CO_2 among various geographic and political entities. There is a distinct variance between industrialized and developing countries, as well as across the economic sectors. This disparity leads to foreboding debates that delay critical action. This chapter calls for an immediate, worldwide shift to a low-carbon economy. It extends this recommendation by suggesting a plethora of opportunities—found in new technologies and practices—that, if applied effectively, will provide solutions.

Chapter 3 initiates with an analysis of the world's energy usage—distinguishing between renewable and nonrenewable sources—and the resulting carbon emissions. Scientific data is presented to illustrate trends in atmospheric CO_2, showing an alarming increase since 1960. Six different carbon-emitting processes are described. A discussion of the carbon cycle is provided in reference to natural processes. In relation to anthropogenic carbon emissions, several examples of oil extraction processes are offered. Thus, carbon-emitting processes are also implicated in a myriad of other environmental issues.

Environmental damage that is closely related to increasing carbon levels includes droughts, forest fires, acidification and much more. Consequently, carbon emissions are reemphasized as a serious environmental issue with the potential to destroy precious ecosystems and landscapes. It discusses the failures and successes of the Kyoto Protocol, particularly outlining the lack of U.S. support for the treaty. In contrast, the Montreal Protocol is presented as an example of successful international cooperation, which has led to the gradual repletion of the ozone layer. This chapter ends with an analysis of environmental frameworks, intended to mitigate global climate change through international agreements.

Chapter 4 deals primarily with government actions that address environmental concerns, specifically carbon emissions and ensuing global climate change. This chapter depicts an eventual transition in legislative acts from those focused on increasing traditional carbon-based energy resources and economic development approaches to those that could reduce environmental damage from irresponsible resource exploitation. Each level of government is discussed in relation to carbon-reducing efforts, from supra-national agreements and regional agreements to state-issued restrictions and policies. Various strategies are examined for their relevance to a specific area, state, or nation—considering population, climate, industry, geography, and other influences. This demonstrates the importance of flexibility in modeling carbon reduction strategies to match a government's particular context.

However, the drawbacks of flexibility are also emphasized in the shortfalls of international agreements and carbon markets. Moreover, the successes and failures at each level of government are presented. This chapter diligently points out the importance of integrating the different levels of government to implement a comprehensive carbon reduction strategy. It advocates a hierarchical structure of overseeing and implementing mitigation policy.

Chapter 5 addresses local carbon reduction policies, as implemented by city governments and often structured by international, national, and regional partnerships. This chapter emphasizes the ability of local solutions to support the longevity of systems and well-being of the populace, especially in terms of carbon reduction. One methodology is for city governments to provide effective leadership for energy consumption and conservation policies. Other measures of achieving carbon reduction refer to the responsible management, reconstruction, and design of energy-efficient infrastructure and urban space. Local policies provide the theoretical basis for local initiatives.

In addition, Chapter 5, "Local Carbon Reduction Policies," emphasizes the importance of the U.S. Mayors Climate Protection Agreement and includes a description of its primary components. Various cities participating in this agreement are discussed throughout the chapter to illustrate the achievements that have accompanied their progressive measures. Overall, the impressive nature of local action is linked to economic incentives, community campaigns, and appropriate urban design approaches that are conducive to sustainable developments and lifestyles. Emphasis is placed on the wide range of options that communities have used in their efforts towards carbon reduction.

Chapter 6 offers an in-depth examination of the building sector, showing its overwhelming proportion of energy use. A description of the different aspects of a building's inner workings and the related energy requirements is provided. This section is especially informative about the numerous opportunities for architects, engineers and building design professionals to create and implement sustainable projects. The development of "green building" as a concept and practice is presented. The ambiguity in meaning and interpretation of "green building" and "green construction" is also discussed.

The series of steps required for verifiable green practices are presented as achievable objectives, though requiring a careful planning process. Consideration of materials, technologies, practices, and designs are all discussed in relation to green building practices; the details, like the process, are quite thorough. Chapter 6 illustrates the working components of a building—water, heating, cooling, electrical systems, building shells, etc.—as part of an integrated, complex system. This chapter shows the range of options available for responsible building, from net positive energy construction to carbon neutral designs, and emphasizes the possible impact on carbon reduction.

Chapter 7 develops a full understanding of the alternative energy sources and their respective potential in the fight against global climate change. A description of various alternative energy sources and the processes involved in their use is included. In particular, the objectives concerning water management are discussed with practical applications for buildings, cities, and nations. Further, the incorporation of energy efficiency measures and technologies into energy systems is shown as a significant contribution to mitigation efforts. Examples of these initiatives are provided throughout the chapter.

Interesting opportunities for alternative energy sources are also described in connection to human activities and designs. Examples discussed in this section include landfill gas extraction and distributed generation technologies. In addition, the controversy surrounding nuclear energy is addressed in a political context. The final paragraphs portray innovative technologies and theories that are still being developed.

Chapter 8 examines the plausibility of how carbon sequestration techniques might be used to lessen atmospheric carbon. Natural carbon sequestration processes are explored in relation to the carbon cycle. In addition, the biological transfers of carbon (as found in the carbon cycle) may also apply to enhancement methodologies or "geo-engineering." Several geo-engineering techniques are described that aim to make natural carbon uptake more effective or increase the carbon cycle's capacity. There are unknown risks associated with attempts to engineer natural enhancements and different hypotheses are discussed to demonstrate the range and level of uncertainty.

Geological storage of carbon emissions is considered with emphasis on the economic and political feasibility of this form of sequestration. The strengths of enhanced oil recovery (EOL) are discussed as well as examples of co-location. The potential for zero emissions coal (ZEC) is very explicit, especially in relation to the U.S. where coal usage is increasingly high. Several innovative, cutting-edge technologies such as algae farms and agricultural methane capture are also included to show future possibilities and comparisons to current methods. Altogether, this chapter offers a new perspective and focuses on how to reduce the carbon content of the atmosphere using extraction processes, in addition to prevention.

Chapter 9 centers on the corporate capacity to incur and support mitigation policies. Corporations may pursue carbon reduction goals through many venues: internal policies, corporate programs, demands on suppliers, etc. Indeed, new market demands and expectations for corporations are identified and closely linked to "clean tech" industry. The automobile industry, utilities companies, service companies, and the banking industry are specifically addressed. Examples of new environmentally-friendly prototypes are included to demonstrate the determination and reformation of the automobile industry. The transition of utility companies from their infamous "see no carbon, hear no carbon, speak no carbon" motto to the application of green technologies and practices is outlined. Similarly, the section illustrates adjustments by highly mobile service companies as well as new options for commercial enterprises. The evolution of "green banking" is provided as an example.

Though depicting an abundance of green options and opportunities, this chapter also presents the obstacles and problems associated with carbon reduction efforts, especially in terms of profit-making ventures and infrastructure capacity. Clearly, much remains to be done by corporations trying to take part in the green revolution. Information tools are important for the success of this endeavor and consequently, this chapter addresses assessment methodologies such as carbon footprints, and financial tools such as carbon credits. In conclusion, this chapter illustrates the importance of corporate involvement and the profusion of opportunities in the green marketplace.

Chapter 10 provides an interesting look at industrial and manufacturing sectors and their preparedness to tackle carbon emission reduction. The ability of these sectors to accomplish carbon goals through energy efficiency is critical for success, especially in the U.S. where industry represents an important portion of the total global opportunity for reducing carbon emissions. Two initiatives of the U.S. Department of Energy are examined as well as the steps required to accomplish and implement them. This chapter discusses the inadequacy and ineffectiveness of "business as usual" approaches. It compares these stagnate strategies to a Pricewaterhouse-Coopers (PWC) report that outlines the implications of population growth on carbon emissions. Several different scenarios in the PWC report are explained.

The metrics of carbon emissions are defined as indicators and included in the analysis of industrial and manufacturing sectors. The steps needed to produce industry-specific indicators are emphasized as important aspects of tracking trends and meeting standards. A proposal for a greenhouse gas intensity index is briefly discussed. Attention is given to the use and development of modeling systems, including examples of "bottom-up" and "top-down" approaches. Finally, the specific application of such measures is connected to various industries. Case studies are analyzed and developing technologies are presented.

Chapter 11 deals primarily with human and organizational development, interlinking the structures and skills available with those needed to accomplish mitigation policies. It specifically addresses the concepts of sustainability and culture in organizational terms. The necessary elements within an organization are explicitly listed and linked to carbon emission reduction strategies. The utilization of internal and external resources is suggested. The experiences of California provide an example of innovative partnerships geared toward researching and developing new technologies. An in-depth analysis of leadership qualities and decision-making processes is depicted through references to Dr. Arun Jhaveri's "Effective Leadership for Sustainable Development in the Public Sector" dissertation. This section outlines various approaches that organizations may take to implement sustainable carbon reduction strategies; in particular, lateral coordination is portrayed as a vital to achieving organization-wide support.

This chapter demonstrates the tendency of organizations to develop and implement sustainability imperatives in a similar manner. Clarity of mission, coordination of teams and lateral communication networks are frequently facilitated by organizational approaches towards carbon reduction. Furthermore, the comprehensive effects of interdisciplinary approaches—spanning various departments within an organization—are shown to be beneficial in redirecting an organization towards mitigation policies.

Chapter 12 illustrates the history of the global environmental movement, offering historical background that is relevant to the management of carbon reduction programs. The importance of transparency is highlighted as a necessary element to securing public support and effective cooperation. The Austin Climate Protection Plan is provided as a potential model for other urban governments. This model is applauded as a truly integrated, interdisciplinary approach. Other examples are also included and show how various approaches in the Austin plan sought a contextual fit—addressing specific needs, requirements, opportunities, obstacles, environments, politics, and investments of each respective entity. This chapter further explains the differences between the top-down and bottom-up approaches, indicating that since the Third Assessment

Report of the IPCC they have become more similar. This chapter concludes by suggesting ways to prevent carbon reduction strategies from failing, and discussing what to do when organizational carbon reduction goals have not been achieved.

Chapter 13 concentrates on the propagation of "cap-and-trade" initiatives throughout the world, clarifying their specific functions and the requirements for participation. The roles of financial institutions include serving as administrative and trading representatives. Specifically, these bodies can control the measurement, pricing, and types of carbon offsets used in carbon trading schemes. The chapter further elaborates on the history of carbon offset transactions and the future potential for carbon markets.

Chapter 13 next considers the financial aspects of carbon capture systems (CCS). While CCS techniques are still in the pre-demonstration stage of the innovation process, they hold enormous potential for abatement programs and are deemed economically advantageous. The relationship between anticipated coal use and CCS is explicitly stated. Examples of CCS projects for coal-fired power generation plants are reviewed.

This chapter provides an approximation of where carbon prices might be in the future. In addition, it indicates how rising carbon prices may be utilized in the procurement of carbon reduction technologies. Effective financial solutions are discussed, ranging from private investments in new technologies to government-issued grants.

Chapter 14 describes a number of successful planning strategies used by public and private sectors, including carbon emission reduction and sustainability models. Incentives for adopting progressive carbon reduction strategies are shared, as are the corresponding results. Furthermore, this chapter mentions critical developments currently in process that might dramatically influence future carbon reduction plans. A comparison of emission trading mechanisms to emission taxes is provided to highlight their respective advantages and concerns.

Transportation issues are also discussed and a list of facts verifies the dramatic influence that mobility has on carbon emissions. Public transportation system improvements are identified for their environmental and social advantages, especially in high density urban landscapes.

Chapter 15 anticipates continual growth of carbon trading systems and carbon markets through economic incentives and corresponding impacts on other carbon-reducing measures. This chapter re-emphasizes the seriousness of CO_2 and greenhouse gas emissions and the high propensity of the U.S. population towards excessive energy use. It also summarizes the different strategies and technologies for reducing CO_2 emissions. Important information tools are described as well as their relevance to garnishing public support for integrated participation in carbon reduction plans. Coherent public disclosure of mitigation strategies are required to amplify effectiveness across sectors, industries, and levels of government. In particular, the importance of coordinating legislation and policies at local, state, and federal levels is justified and encouraged. Furthermore, new possibilities are outlined as well as their anticipated impacts on the design of infrastructure. The chapter advocates integrated, cohesive carbon reduction strategies and offers evidence to support the need for progressive measures.

The crossroad that is upon us creates a multitude of dilemmas and exciting opportunities. One dilemma is that failure to take action today to reduce carbon emissions exacerbates the difficulty of reducing carbon levels in the future. To solve this problem we must change our agendas, rewrite our policies, institute new programs and implement new technological solutions. There is much work to be accomplished and we have only just begun to recognize solutions. *Carbon Reduction Policies: Strategies and Technologies* delves deeply into the solutions that are necessary to achieve the goal of reducing the amount of carbon being released into the atmosphere. Time is of the essence.

Chapter 1

The Importance of Reducing Carbon Emissions

"If further global warming reaches 2 or 3 degrees Celsius, we will likely see changes that (will) make Earth a different planet than the one we know. The last time it was that warm was in the middle of the Pliocene, about 3 million years ago, when sea levels were estimated to have been about 25 meters (80 feet) higher than today."

James Hansen
NASA's Goddard Institute for Space Studies
New York (2006)

We know now what the problem is. We have been successfully tracking gradual increases in atmospheric carbon emissions. We know this is serious. The importance of finding a solution to reducing carbon emissions cannot be understated. The emissions of atmospheric carbon, generated by man's activities, are exacerbating environmental changes on a global scale. In the past, we seem to have made conscious choices to ignore potential consequences. Today, the political and social structures we have established are developing only embryonic policies and taking only feeble action. Our initiatives often seemed mismatched and the actions marginal. We have pushed aside our stewardship responsibilities and failed to bring our technical and economic resources to bear on the task of providing more viable solutions.

Consequently, new environmental problems have surfaced that emphasize the ultimate fragility of our planet's ecosystems. Due to our unresponsive choices, climate changes are occurring at a rapidly accelerated pace. Our common future is becoming grim. No matter what actions we take today, the consequences of increasing carbon levels will continue to plague the earth for generations to come. Never before have human-induced environmental changes had such a drastic impact. At this crossroads, nothing less than the sustainability of life on our planet is at stake.

The results of our past insensitivity to the environment are becoming more and more noticeable, the consequences less and less dismissible. With the acceptance that an over-abundance of atmospheric carbon dioxide (CO_2) is impacting Earth's ecosystems and can be a planet-destroying greenhouse gas, a new age is dawning. The economies of the Hydrocarbon Age are at an impasse. Indeed, the availability of some fossil fuels is reaching a tipping point. We are changing our landscapes—terra-forming them to meet the needs of agriculture, industry, urban expansion and regional transportation networks—as a result of economic development and technological advancements. Many of our cities have evolved to become "post-industrial" and are unlike anything that our forbearers might have imagined. Other cities, like some in Mexico, China and India, are reeling from explosive population growth, poor development choices and environmental damage. Countries seeking economic and political control over carbon-based energy resources are at times exerting their military muscle, threatening and waging regional conflicts. In this climate of uncertainty, the impacts of each country's policies and economic systems are intensifying the problems associated with carbon emissions rather than resolving them. However, the political dialogue is shifting from outright rejection of the existence of carbon-induced climate problems to the development of strategies and legislation to mitigate them.

At this point in history—when creative solutions are needed—the backdrop of conflicting influences, political stalemates and economic turmoil makes long-term solutions difficult to implement. Energy, environment and economy are intimately linked. For example, energy-consuming systems account for 95% of man-made CO_2 emissions (Mega 2005:60). According to Kenneth Cohen, Vice President of Public Affairs for Exxon Mobile Corporation, the real issue is "how to provide the energy needed to improve global living standards while also reducing greenhouse gas emissions."[1] The seminal inconvenient truth is that a consensus regarding how to proceed remains elusive. Regardless, governments are creating policies, corporations are reconstructing strategic plans, and institutions are redefining their missions. Agendas are changing and new programs are being launched and implemented. There is an evolving consensus on the horizon; one that will change how we prioritize our efforts to become more sustainable. The time has come to refocus our resources towards finding solutions that reduce or eliminate carbon emissions.

Mankind's misuse of carbon-based fuels has created a new climate of uncertainty. While there are both natural and man-made sources of atmospheric carbon, natural endowments that reduce atmospheric carbon levels are under attack by our activities. Formed over hundreds of millions of years, carbon compounds were stored by natural processes that worked very slowly. These processes gradually reduced atmospheric concentrations by storing the carbon underground and beneath the seas. There are currently no substitutes for these natural processes that have gradually and effectively sequestered massive quantities of carbon for eons.

INCREASED USE OF CARBON-BASED FUELS

Today, the carbon compounds that are being released into the atmosphere come from many sources, including primary combustion of fuels such as coal, natural gas and oil. The use of coal represents more than half of the CO_2 being released into the atmosphere in the U.S. Despite its high levels of CO_2 emissions and its low thermal energy conversion efficiency (approximately 37%), coal usage will continue to increase and play an important role in global energy supply.

What is happening is that we are increasing our use of carbon-based fuels—and the atmospheric concentrations of carbon dioxide are also increasing. A consensus has evolved that global climate change is a result. The debate is over. We know that carbon emissions are directly linked to the use of fossil fuels and they are responsible for global climate change. Mankind's activities are exacerbating this problem and creating climatic disruption, loss of biodiversity and economic uncertainty. Despite increased awareness, efforts to address the impacts of climate changes are in their infancy; they have yielded little impact on global emissions. In fact, carbon emissions worldwide continue to increase. This is due to the growing quantities of fossil fuels being extracted and how they are used in combustion processes.

The concentration of carbon in fuels varies, as does the amounts of carbon released into the atmosphere during the combustion processes. One way to reduce CO_2 emissions is to become less dependent on fuels with high carbon content. Efforts should be made to replace such fuels with those that generate less carbon. Table 1-1 below provides a list of fuels and their carbon content.

Table 1-1. Carbon Intensity of Fuels[2]

Fuel Type	CO_2 Emissions Produced (kilograms)
1 gallon of gasoline	8.9
1 liter of gasoline	2.4
1 gallon of diesel fuel	10.2
1 liter of diesel fuel	2.7
1 short ton of bituminous coal	2.24
1 short ton of lignite coal	1.27
1,000 cubic feet of natural gas	54.7
1 cubic meter of natural gas	1.9
1 kWh of electricity	0.61

There are two primary ways to manage atmospheric levels of carbon. Emissions can be reduced or carbon can be extracted from the atmosphere. Like other atmospheric pollutants, once emitted into the atmosphere, carbon is diffused and both difficult and expensive to extract. The technologies available to remove carbon from the atmosphere are commercially unproven and many are untested. Thus, reducing the quantities of carbon compounds prior to their release into the atmosphere holds far greater promise.

Our habits over the last 60 years have contributed to the dire situation we find ourselves in today. Many of the adverse environmental consequences we face are caused by increased atmospheric carbon concentrations and their impact has only recently become understood. Why is this? As with serious environmental problems encountered in the past (e.g., natural resource depletion or water pollution), there is a time lag before the environmental impacts of human actions become apparent. Environmental changes must be meticulously observed before the causes and effects are assessed. Theories typically evolve as to why the changes are occurring. Their projected impact must be considered. To do this, the scientific community must make comparative measurements of before and after conditions, link the environmental changes to the causes, and consider an assortment of remedial actions.

Prior to suggesting approaches to mitigation, a wide array of intervention methodologies and technologies must be developed, tested and deployed. After a period of societal denial and disbelief a political consensus must emerge before public resources can be used. This is a time-intensive and laborious set of activities. Meanwhile, our present infrastructure continues to operate, exacerbating any problems it has created.

To complicate matters, seldom is there a single solution that will provide a cure. The "smoking gun" is oftentimes more readily identified than the elusive "silver bullet"—and none may exist. In fact, scientific investigations often conclude that a set of customized mitigation approaches is necessary. Such contextual responses vary in their identification of the source of the problems, local circumstances, economic costs and the scope of the solution. In some cases, recommended technologies have unintended results that are not necessarily favorable. For example, in 2008 24% of the corn grown in the U.S. was harvested to manufacture ethanol and used as an alternative fuel. The use of corn for fuel production is expected to increase to 32% of the total crop in 2009. This has increased pressure on corn prices and has stressed food supplies.

A scientific consensus regarding the environmental damage caused by carbon emissions, has only recently become available. Increasingly, the deleterious consequences of carbon-induced climate change have become more apparent. The consensus is that carbon emissions are contributing to possibly irreversible changes in the world climate. The time of inaction has passed and the time of initiative is upon us. We must take action now or our common future is in peril.

CARBON EMISSIONS

Global carbon emissions from man-made sources more than tripled from 1950 to 2000 and they continue to increase. From 1990 to 1999, global emissions increased at a rate of 1.1% annually, jumping to 3% annually afterwards despite the advent of the Kyoto Treaty.[3] Carbon dioxide emissions (CO_2e) increased to 28.2 billion metric tons in 2005.[4] Moreover, projections suggest that CO_2e will climb to 33.9 billion metric tons in 2015 and to 42.9 billion metric tons in 2030.[5] Increases of such magnitude are unprecedented. The potential impacts to the earth's ecosystems are unknown. It is likely that further damage to the earth's climate will occur. This damage will have unforeseen consequences and impact our economies, our health, our resources and our settlements. These changes are likely to occur so quickly that we may be unable to adjust rapidly enough to avoid global catastrophe.

Greenhouse gas emissions in the U.S. in 2006 have been estimated to be 7.1 gigatons.[6] Of this total, approximately 6.0 gigatons was emitted as CO_2.[7] In addition, 0.5 gigatons were emitted as methane (CH_4), a potent greenhouse gas 25 times more damaging than CO_2. Methane is used as a combustion fuel for electrical production and escapes into the atmosphere from fermentation, landfills, coal mining operations, manure, natural gas systems and other sources. Since burning methane releases less CO_2 per unit of heat generated than other hydrocarbon fuels, it is seldom the fuel of choice. With a half-life of seven years, methane in the atmosphere oxidizes and produces CO_2 and water.[8] In comparison to CO_2, each methane molecule has a relatively large global warming impact that diminishes in a relatively short period of time. Methane tends to concentrate in the stratosphere and in tropical regions. The amount of methane present in the atmosphere has increased from 700 parts per billion (ppb) in 1750 to 1,745 ppb in 1998, with more than half due to human activity.[9] For these reasons, efforts to mitigate atmospheric carbon concentrations must include preventing the release of methane gas.

The distribution of carbon emissions from the world's countries is unevenly distributed. The U.S. emits CO_2 at more than twice the per capita rates of Germany, Japan, or Russia. In more populous, industrializing countries the rates of increase are becoming unmanageable. In 1990, China and India combined for only 13% of world emissions yet their share increased to 22% by 2004 and is anticipated to further increase to 31% by 2030.[10] China constructs the "equivalent of two midsize coal-fired power plants each week—adding a capacity comparable to the entire UK power grid each year."[11] In 2006, China surpassed the U.S. as the world's number one carbon dioxide polluter.

CO_2 emissions are often concentrated in or near urban areas. Canadian cities produce roughly 20 tons of CO_2 per capita per year while residents of Amsterdam produce only 10 tons annually on a per capita basis (Roseland 1998:16). CO_2 emissions can also be considered as an average based on population which includes both urban and non-urban areas. There is wide variability in emissions rates. Table 1-2

Table 1-2. National Levels of Fossil Fuel Consumption

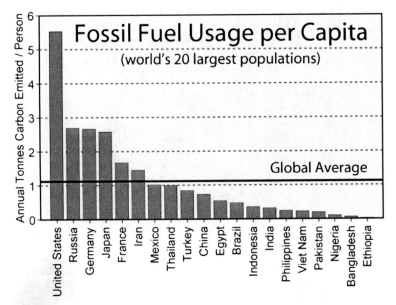

Source: Carbon Dioxide Gallery, developed by Rohde, R. from publicly available data. http://www.globalwarmingart.com/wiki/Image:Fossil_Fuel_Usage_png, accessed 21 September 2007.

uses national per capita emissions based on fossil fuel consumption as a measure. The U.S. ranks highest with an index of roughly 5.5 metric tons emitted per person annually.

Within countries, greenhouse gas emissions vary widely across economic sectors. The production of greenhouse gases due to power generation, industrial processes and transportation account for over 52% of the world's total greenhouse gases emissions. Strategies to reduce carbon emissions will impact all sectors of the economy and costs are likely to be unevenly distributed... regardless of the strategies employed. Therefore, policies that focus on reducing carbon emissions must be broadly based. A consensus must be reached. Programs must be implemented to achieve the goals established by the policies that are adopted. While policies and programs are replicable, they must be overarching in their principles, consistent in their goals, and locally adaptable. The technologies employed must be appropriate, feasible and economically viable.

Figure 1-1. Annual Greenhouse Gas Emission by Sector

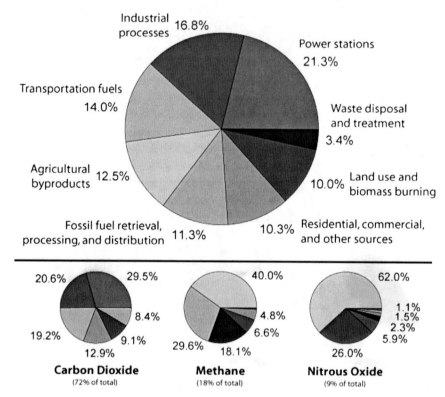

Source: Earth System Research Laboratory, Global Monitoring Division, developed by Rohde, R. from publicly available data. http://www.globalwarmingart.com/images/e/e0/Greenhouse_Gas_by_Sector.png, accessed 21 September 2007.

THE COSTS OF REDUCING U.S. CARBON EMISSIONS

The cost of reducing U.S. greenhouse gases is difficult to estimate. The projections range from almost nothing to hundreds of billions, if not trillions, of dollars. According to a recent report entitled *Reducing U.S. Greenhouse Gas Emissions: How Much at What Cost?* by McKinsey and Company, annual greenhouse gas emissions in the U.S. are projected to increase from 7.2 gigatons[12] of CO_2 equivalence in 2005 to 9.7 gigatons in 2030 if no remedial actions are undertaken.[13] This report identifies the primary causes for the projected growth of U.S. carbon emissions:

1) The anticipated expansion of the economy.
2) Growth in the use of energy by buildings, appliances and transportation due to a projected population growth increase of 70 million.
3) The continued reliance on carbon-based electrical power generation from the construction of new coal-fired power plants that lack carbon capture and storage (CCS) technology.
4) A gradual decline in the ability of U.S. forests and agricultural lands to absorb carbon, forecasted to decrease from 1.1 gigatons in 2005 to 1.0 gigatons in 2030.

The McKinsey and Company report identifies three case projections (low range, mid-range and high-range) that consider abatement opportunities such as the greater use of coal as an energy source using CCS, expanded use of nuclear power, renewable energy and biofuels, along with vehicle efficiency improvements and energy efficiency upgrades for buildings. McKinsey and Company's carbon abatement projections establish a cost of $50 per ton or less. However, vehicle efficiency improvements typically require investments greater than $50 per ton.

Table 1-3. Economic Efficiency of Fossil Fuel Consumption

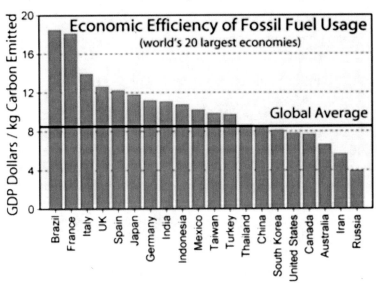

Source: Carbon Dioxide Gallery, developed by Rohde, R. from publicly available data. http://www.globalwarmingart.com/wiki/Image:Fossil_Fuel_Efficiency_png, accessed 21 September 2007.

POTENTIAL FOR REDUCING CARBON EMISSIONS

There remain opportunities to reduce carbon emissions. The potential in the U.S. can be measured in how policies, programs and technologies are effectively synchronized to implement this goal. There must be a focus on managing broadly based initiatives that successfully reduce carbon emissions. However, the pressures of population growth and development temper the belief that we can reduce carbon emissions simply by initiating additional energy conservation and efficiency improvements.

The study by McKinsey and Company evaluates five categories of technologies that can be employed to reduce carbon emissions and estimates their potential impact. These categories are:[14]

- Reducing the carbon intensity from electric power production by using alternative energy and CCS technologies (800-1,570 megatons).

- Improved energy efficiency in buildings and appliances (710-870 megatons).

- Implementing carbon reduction opportunities in the industrial sector (620-770 megatons).

- Expanding natural carbon sinks to capture and store carbon (440-590 megatons).

- Increasing vehicular efficiency and using less carbon-intensive fuels (340-660 megatons).

These approaches, if considered as additive, offer an estimated total reduction in carbon emissions ranging from 2,910 to 4,460 megatons, or a reduction from the base year (2005) emissions ranging from 40 to 62%. While these reductions in emissions are possible, infrastructure efficiency improvements will be necessary. When implemented, such initiatives would place the U.S. at a per capita carbon emission level close to that of Germany or Japan.

While the U.S. is improving the efficiency of vehicles and new buildings, it is clear that the country needs to direct more resources and efforts towards reducing emissions from electrical energy production and improving the efficiency of existing buildings and appliances—areas where the greatest reductions in carbon emissions are available. The need to expand the U.S. economy and support a growing population makes the implementation of carbon emission mitigation programs more challenging. However, there are ways to reduce the impact of carbon emissions without adversely impacting the economy. Reducing electrical demand is a prime example. Projections indicate that improvements in building and appliance energy efficiency combined with industrial sector initiatives could offset 85% of the incremental demand for electricity through 2030.[15] This scenario is certainly possible, yet it is much more likely that the increased need for electricity through 2030 will be met by increasing the number of coal-fired and nuclear power plants.

TRANSPORTATION SYSTEMS

Energy production and transportation systems are generating pollution, and atmospheric carbon levels are increasing in tandem. Emissions from transportation-related sources now account for roughly one-fifth of total carbon emissions and are increasing rapidly in developing countries such as China and India (Kaihla 2007:70). Some countries, such as Brazil and France, are able to operate their economies using much less

High-speed Eurostar Trains in Rome, Italy

Freight Rail Yard East of Kansas City, Kansas

carbon than countries such as the U.S., Canada and Australia. Vehicles in Brazil use more ethanol than vehicles in other countries. France has high rates of passenger train usage and relies on atomic power for electrical production. We do have transportation options. In the U.S. and elsewhere in the world, moving freight and people by rail is not only more cost-effective, but also less carbon intensive. High-speed passenger rail services in Europe have been expanded and offer a wide range of inter-city travel options.

SUMMARY

Carbon emissions generated by man's activities are contributing to potentially irreversible changes in the world climate. The scientific consensus is that global climate change is one result. Nothing less than the sustainability of life on our planet is at stake. Natural processes have effectively stored large quantities of carbon for eons and there are currently no substitutes for these natural processes. The carbon compounds that are being exhausted into the atmosphere come from many sources—especially fossil fuels. The Hydrocarbon Age is at an impasse. We must find ways to use carbon-based fuels more efficiently and effectively.

We know that energy, environment and economy are linked; yet a consensus regarding how to effectively deal with this relationship remains elusive. There is no single, quick fix solution since greenhouse gas emissions vary widely across countries and economic sectors. Long-term solutions that reduce carbon emissions are needed yet they are costly and difficult to implement. Policies that focus on reducing carbon emissions must be broadly based. Programs must be implemented to achieve the goals established by those policies that we choose to adopt. Technologies that are directed towards the reduction of carbon emissions must be adaptable and economically viable.

Solutions include expanding natural carbon sinks, improving the energy efficiency of vehicles, upgrading buildings and facilities, reducing emissions caused by electrical energy production and using more efficient appliances. The potential can be measured in how policies, programs and technologies are effectively synchronized to reduce carbon emissions.

Footnotes

1. Cohen, K. (2007, 3 September). Letters. *Newsweek.* p. 18.
2. McKinsey & Company (2007, December). *Reducing U.S. greenhouse gas emissions: how much at what cost?* U.S. Greenhouse Gas Abatement Mapping Initiative. p. 74.
3. Friedman T. (2008). *Hot, flat and crowded.* Farrar, Straus and Giroux. p. 214.
4. Energy Information Administration (2007, September 18). *International energy annual 2005.* http://www.eia.doe.gov/pub/international/iealf/tableh1co2.xls, accessed 3 November 2007.
5. Energy Information Administration (2007, May). *International energy outlook 2007.* http://www.eia.doe.gov/oiaf/ieo/emissions.html, accessed 3 November 2007.
6. U.S. EPA (2008, February). *Inventory of U.S. greenhouse gas emissions and sinks1990-2006.* Public Review Draft.
7. *Ibid.*
8. Wikipedia. *Methane.* http://en.wikipedia.org/wiki/Methane, accessed 8 March 2008. The chemical formula for this process is $CH_4 + 2O_2 -> CO_2 + 2H_2O$.
9. *Ibid.*
10. Energy Information Administration (2007, May). *International energy outlook 2007.* http://www.eia.doe.gov/oiaf/ieo/emissions.html, accessed 3 November 2007.
11. *National Geographic* (2006 May). p. 144.
12. One gigaton is equal to one billion (10^6) tons.
13. McKinsey & Company (2007, December). *Reducing U.S. greenhouse gas emissions: How much at what cost?* U.S. greenhouse gas abatement mapping initiative, Executive report. http://www.mckinsey.com/clientservice/ccsi/greenhousegas.asp, accessed 3 February 2008.
14. *Ibid.*, pages xiii-xv.
15. *Ibid.*

Chapter 2

The Problems of Climate Change

"*Now comes the threat of climate crisis—a threat that is real, rising, imminent and universal. Once again, it is the 11th hour. The penalties for ignoring this challenge are immense and growing, and at some near point would be unsustainable and unrecoverable. For now, we still have the power to choose our fate, and the remaining question is only this: Have we the will to act vigorously and in time, or will we remain imprisoned by a dangerous illusion?*"

*Al Gore, Former U.S. Vice President and
2007 Nobel Peace Prize recipient
in his acceptance speech,
"Making Peace with the Planet."*

THE BASICS OF CLIMATE CHANGE

According to many world-renowned scientists, economists, political leaders and technical documents, it is clearly evident that the surface temperatures on Earth are warming at a pace that signals a decisive shift in the global climate. Scientists warn that current CO_2 emissions must be reduced by at least half over the next 50 years in order to avert a disastrous climate change.[1]

Before the Industrial Revolution, Earth's atmosphere contained about 280 parts per million (ppm) of CO_2. The atmosphere equated to a global average temperature of about 14°C (57°F)—a reasonably acceptable and habitable environment. The Industrial Revolution and the use of coal for steam power was in part responsible for increases in atmospheric CO_2 concentrations at a time when these increases were not being monitored. When humans started burning more coal, gas and oil for heat and power, the amount of atmospheric CO_2 increased. By the time atmospheric CO_2 measurements were first established in the late 1950s, emissions had already reached the 315-ppm level. Currently, CO_2 emissions have reached 380 ppm; they are increasing by approximately 2 ppm each year. Though the rate seems trivial, the additional CO_2e traps enough extra heat to warm the planet significantly. Earth's temperature has increased more than 1°F in the past 30 years.[2]

Although it is impossible to precisely predict the consequences of further CO_2e, there are indications in geological evidence and science-based models. The warming as observed to date has triggered glacial retreat and Arctic ice melt. Climate change has distorted seasons, altered rainfall patterns and caused sea levels to rise. Based on the current scientific data demonstrating global climate change, 450 ppm of CO_2 has been repeatedly and strongly recommended as a threshold. Beyond 450 ppm, the earth's environment becomes vulnerable to irreversible, detrimental impacts.[3] Averting the trend towards higher global temperatures requires an urgent, worldwide shift to low-carbon economies; this initiative should encompass innovative carbon reduction technologies. Earth's atmosphere is expected to reach this threshold within 35 years, if the 2 ppm annual increase in CO_2e continues unabated. Furthermore, this scenario excludes the impact of other greenhouse gases (GHGs) such as methane and nitrous oxide, which also contribute to global climate change.

Thus far, the European countries and Japan have begun to trim carbon emissions, though they may not meet their modest targets. Meanwhile, U.S. carbon emissions—a quarter of the world's total—continue to rise steadily, with an estimated increase of 20% more in 2020 compared to 2000. China and India are now generating increasingly large quantities of CO_2 as well. On a per capita basis, these countries emit far less CO_2 than the U.S. However, their populations are so large and their growth is so rapid that lowering worldwide CO_2 emissions will be challenging. Preventing a global catastrophe will require rapid, dramatic cuts in CO_2 emissions by technologically advanced countries. Countries such as the U.S., European Union, Japan and others must take the lead in reforms. This can enable developing countries to power their emerging economies without a heavy dependency on CO_2-emitting coal and other fossil fuels like gas and oil. Mitigation represents the fundamental concern of carbon reduction strategies.

CONCEPTUAL FRAMEWORK

An overwhelming scientific consensus has emerged that human beings are causing significant changes in global climate systems. The challenge presented by climate change is daunting and merits urgent attention. Current estimates indicate that global GHG emissions will need to be reduced 80% below the current levels by 2050 if catastrophic warming is to be avoided—that is, to keep the increase over the current average temperature to "only" 2 degrees Celsius (C) or 3.6 degrees Fahrenheit (F) by 2100.[4] To meet this goal, developed or industrialized nations must reduce their emissions by 95%. The various scenarios projected by the UNFCCC (United Nations Framework Convention on Climate Change) predict a 1.8 to 4.0°C (3.2 to 7.2°F) rise by the century's end (2100), with an average best estimate of 2.8°C (5°F).

Once released, CO_2 remains in the atmosphere for up to 100 years or more.[5] It is the existing stock of carbon in the atmosphere that is driving climate change, while the annual flow of emissions exacerbates the associated problems. If not addressed, climate change will result in a host of negative outcomes, including severe, long-term, destructive economic impacts.[6] The anticipated effects of climate change pose a serious threat to human societies, much more so than issues like pollution (smog, for instance) or congestion (such as traffic jams). Hence, solutions and preventative measures are desperately needed.

The conceptual framework involves finding solutions that will address climate change and reduce GHG emissions, particularly those that are causing increases in atmospheric CO_2 concentrations. Therefore, the question is not whether GHG and CO_2 emissions need to be trimmed and monitored. The real issues are to what extent, with what strategies, and when, should the current GHG and CO_2 emissions be cut—considering both the short-term and long-term effects.

Finding ways to significantly slow the onslaught of global climate change is an important task for engineers, scientists, corporations, communities, governments and decision makers around the world. Ways to mitigate climate change might involve the prompt implementation of innovative policies, strategies and technologies associated with energy efficiency, renewables, sustainable/green buildings and alternative-fueled transportation vehicles and systems. Mitigating climate change will also require a worldwide shift to a low-carbon economic platform. This platform involves implementing a radical reformation of policies, programs and technologies. Among the many technologies and strategies for "low carbon solutions" are the following:

1) Emissions Limit/Caps
2) Emissions Trading/Market-based, Least cost approach

3) Energy Efficiency
4) Renewable Energy Sources
5) Fuel Switching (to lower carbon-content fuels)
6) Offsets/GHG Credits
7) Sequestration
8) Sustainable Green Buildings—Design/Construction/Renovation
9) Waste Management/Recycling
10) Improving the Fuel Efficiency of Vehicles (e.g., Hybrid Vehicles)
11) Tax Incentives, Rebates, and Alternative/Creative Financing
12) Education, Outreach, Awareness, Training, and Communications

THE PROBLEM OF CARBON DIOXIDE

The earth's atmosphere traps solar heat in a manner that can be compared to a greenhouse. Greenhouse Earth is hospitable to life, retaining enough of the sun's heat to allow plants and animals to exist. This natural climate control system depends on the trace presence of certain atmospheric gases—most importantly CO_2—to trap the sun's radiation. But the system's stability has been jolted. The burning of carbon-laden fossil fuels (primarily oil, gas and coal) has hiked atmospheric CO_2 to levels unprecedented during human history. Greenhouse Earth is growing warmer. How much warmer it gets, depends on the human response. This is the most complex and difficult problem that the world has faced for decades, if not for generations, requiring unprecedented and urgent leadership, innovations, resources and commitment. Failure to neutralize the accelerating increase of greenhouse gas emissions (especially methane and carbon dioxide) by 2050 is irresponsible.[7]

The energy sources that propel modern industrial societies have also seeded the climate change crisis. Fossil fuels—oil, coal, and natural gas—account for nearly 80% of the added CO_2 that traps more heat in the atmosphere. Much of the rest of the CO_2 being emitted comes from land-use changes, primarily in tropical forests, including slash-and-burn agriculture and timber harvesting.

During the past century, the pace of warming has accelerated. Eleven of the twelve warmest years on record have occurred since 1995. Because the earth's surface heats up more slowly than the atmosphere, temperatures have not yet caught up with the escalating concentration of CO_2—a very serious problem, particularly if we fail to confront and correct the accelerating rise of CO_2 emissions from human activities. In the past, the bulk of greenhouse gas emissions have been generated by the developed-world, including countries such as the U.S., Japan, and those in the EU. Today, rapidly industrializing countries such as China are emitting ever greater amounts of CO_2. China recently eclipsed the U.S. as the leading CO_2 emitter. Yet, on a per capita basis, its emissions are still quite low. In contrast, the countries of Africa are responsible for roughly 3% of the CO_2 that is present in the atmosphere.

Before the developing nations like China, India, Brazil, and Mexico become dangerously higher emitters of CO_2, industrialized countries should share innovative, cost-effective energy and GHG emissions reduction technologies. The hope of such collaborative efforts is to reduce atmospheric CO_2 to levels that protect the earth's natural and man-made environments from the threat of catastrophic climate change. With more international collaboration and technology transfer, the overall global increase in CO_2 can be contained.

The problem is that atmospheric levels of CO_2 are increasing due to man's activities. The importance of addressing this global environmental problem cannot be understated. Mitigation is required to prevent further damage to global ecosystems. These actions require the development of new policies and programs, allocation of economic resources and the implementation of targeted mitigation technologies.

THE LINKS BETWEEN FOSSIL FUELS AND CARBON EMISSIONS

Fossil fuels generate CO_2 emissions when used in combustion processes. The relationship between the use of fossil fuels like oil, coal, and natural gas, and the generation of greenhouse gases is undeniable. When we burn or consume these carbon-rich energy sources for our heating, cooling, lighting, transportation, electrical generation, and industrial processes, we inevitably cause environmental damage.

A group of Princeton University researchers and scientists cautioned us in 2007 to cut the existing CO_2 emissions by half over the next 50 years, in order to prevent global climate change.[8] Some of the recommended strategies include, but are not limited to:

Efficiency and Conservation
   1) Improving the fuel economy of the two billion cars expected on

the road by 2057 to 25.5 (kilometers per liter) (60 miles per gallon) or from the current 12.75 km/l (30 mpg);

2) Reducing the current average building energy consumption by additional 25% within a decade;

3) Improving coal-fired plant efficiency to 60% from the current 40% within a decade;

Carbon Capture and Storage

1) Introducing sequestering systems to capture and store CO_2 underground at 800 large coal-fired plants and/or 1,600 natural gas fired plants within the next 10 to 15 years;

Low Carbon Fuels

1) Replacing 1,400 large coal-fired plants with natural gas-fired plants (low carbon fuels) within the next five to 10 years;

2) Increasing wind-generating power to 25 times the current capacity, as quickly and cost-effectively as possible; and

3) Increasing the use of solar power to 700 times the current capacity within the next 10 years.

National and international projects and programs are achievable and within the reach of CO_2 emitting nations, using available and emerging technologies. Solutions will entail improving best practices, providing creative financing, and sustained political commitment. Innovative public-private partnerships, grass-roots public involvement, as well as carbon-free or carbon-less energy resources are necessary. Implementing new policies, programs and projects will help reduce future carbon emissions and create new economic opportunities.

THE COSTS OF ADJUSTMENT STRATEGIES

The ability of mankind to adjust to increases in global temperatures is limited. The costs of moving settlements and infrastructure from ocean coastlines is indeterminable and beyond the capacity of the world's economies. Geo-engineering technologies that would attempt to lower the earth's temperatures are unproven and likely more costly than shifting to the use of low carbon energy resources. Ideas that have been suggested, such as pumping seawater into the atmosphere or cloud seeding to increase the reflectivity of clouds, and blanketing the world's glaciers to prevent their melting, are adjustment strategies that seem futuristic and costly. Implementing such strategies is perilous and fraught with uncertainties and unforeseen complications.

SUMMARY

The warming of the earth's surface temperatures has been related to increases in carbon emissions since the Industrial Revolution. With the development and growth of fossil fuel-based economies, the original 280-ppm level of CO_2e began to rise at an unprecedented rate. There is great uncertainty surrounding the future impact of global climate change. It is very difficult for scientists and researchers to model and predict the onslaught of global climate changes due to increasing CO_2 emissions because of the complexity of Earth's climate and ecosystems. Yet any uncertainties regarding the nature and degree of global climate change do not negate the urgency of our situation nor the scientific evidence supporting the need for mitigation policies.

This chapter defines a precipice and creates a sense of urgency as the world rushes towards its edge. It identifies an atmospheric concentration threshold of 450 ppm of CO_2, beyond which the environment is likely to endure irreversible, detrimental impacts. At present rates of increases, this threshold will be breached within 35 years.

Averting such a global catastrophe requires rapid, sustained and dramatic cuts in CO_2 emissions by industrialized countries plus large-scale technology investments in developing nations. This chapter calls for a worldwide shift to a low-carbon economy, offering a variety of tools—innovative carbon reduction technologies—that will quell the rise in GHG emissions. Averting irreparable climate change will require a global effort, in conjunction with organizations such as the United Nations and the International Panel on Climate Change (IPCC).

More difficult questions surface about the specifics of mitigation policy. Who should bear the brunt of the reductions? Should the United States take the lead because of its overall contribution to GHG emissions and advanced technology and industry? Should China because of its projected GHG emissions, which now surpass those of other nations? Should developing nations have to cut GHG emissions along with developed nations or should they be allowed to focus on their economic development? *Indeed, this chapter presents the conceptual framework of the mitigation policy debate: What are the predicted environmental consequences? How can mitigation of GHGs be accomplished, with what strategies, by whom, and what technologies should we use to reduce GHG and CO$_2$ emissions?* Chapter 3 initiates answers to these questions by exploring the environmental impacts of carbon emissions.

Endnotes

1. McKibben, B. (2007, October). Confronting carbon: carbon's new math. *National Geographic*. Researched by Socolow, R. and Pacala, S.
2. *Ibid.*
3. *Ibid.*
4. Redeko, B. (2007). *International Leadership Association*.
5. Office of Climate Change (October 2006). *The Stern Review on the economics of climate change*.
6. Fourth Assessment Report. (2007). Intergovernmental Panel on Climate Change.
7. Climate connections. (2007, October). *National Geographic*
8. McKibben, B. (2007, October). Confronting carbon: carbon's new math. *National Geographic*. Researched by Socolow, R. and Pacala, S. p. 33-37.

Chapter 3

The Environmental Impacts of Carbon Emissions

"Every step we take to reduce energy consumption, whether by installing more efficient lighting or encouraging alternative cooling systems, benefits the environment."

Kevin Burke
Chairman and CEO of Con Edison (2007)[1]

Categories of the energy we use are divided unevenly into non-renewable sources (typically carbon-based energy sources such as coal, oil, and oil shale), and renewable sources, such as wind power, solar, geothermal, and gravitational water sources (e.g., hydroelectric, tidal, and water currents). Most carbon-based energy forms being used as primary fuels are fossilized biomass and can be naturally renewed, but not during the lifetime of humanity. The geophysical and biochemical processes involved require hundreds of millions of years for the renewal process to be effective. Much of the solar energy captured and stored in the Earth by fossilized hydrocarbon processes that are being extracted and consumed today, date from the Paleozoic period, roughly 600 million years ago—predating mankind's existence. As a result, renewable energy sources can be further categorized as *sustainable*, while most nonrenewable energy sources are *potentially unsustainable*, and potentially exhaustible. While there is movement toward greater use of renewable energy, hydrocarbon-based energy consumption is also increasing. Fuels sources such as those derived from waste streams (e.g., sawdust, landfill gas, animal wastes, etc.) are more difficult to classify as they are carbon-based fuels yet are rapidly renewable if the waste streams are maintained. These are typically classified as biomass fuels and considered to be renewable.

Ice core data provide a record of the levels of carbon in the Earth's atmosphere. Over the last 800,000 years, the amounts of atmospheric carbon have varied from 180 ppm to a pre-industrial level of 280 ppm. The concentrations of CO_2 have since increased to 385 ppm (approximately 800 gigatons) and continue to augment. The 1994 level of 358 ppm exceeded any verifiable levels of CO_2 over the last 220,000 years.[2]

The amount of CO_2 released into the atmosphere annually by human activities is estimated to be approximately 27 billion metric tons (30 billion tons).[3] This is only a small fraction of the 130 to 255 million metric tons (145 to 255 million tons) that are typically released by volcanic eruptions.[4] Table 3.1 shows the recent history of measured variations in atmospheric CO_2 concentrations since 1960.

Table 3-1. Atmospheric Levels of CO_2

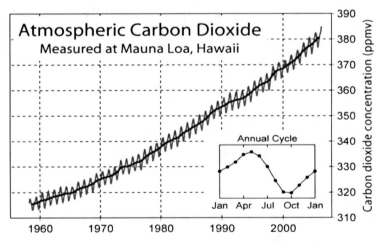

Source: Carbon Dioxide Gallery, developed by Rohde, R. from published data. www.globalwarmingart.com/wiki/Image:Mauna_Loa_Carbon_Dioxide_png, accessed 21 September 2007.

SOURCES OF ATMOSPHERIC CARBON DIOXIDE

Though concentrations of atmospheric carbon dioxide are increasing, the sources can be readily identified as they result mainly from six processes:[5]

1) As a byproduct in ammonia and hydrogen plants, where methane is converted to CO_2;
2) From combustion of carbonaceous fuels;
3) As a byproduct of fermentation;
4) From thermal decomposition of calcium carbonate ($CaCO_3$);
5) As a byproduct of sodium phosphate manufacture;
6) Directly from natural CO_2 gas wells.

Carbon dioxide is released into the atmosphere by the combustion of carbon-containing fuels and petroleum-based distillates. Carbon-containing fuels include oil, wood, coal and natural gas (e.g., methane). Petroleum distillates that generate carbon include gasoline, kerosene, propane and diesel fuels. The chemical reaction between methane and oxygen that creates CO_2 and water is given below.

$$CH_4 + 2\ O_2 \rightarrow CO_2 + 2\ H_2O$$

Yeast produces carbon dioxide and ethanol, by natural causes and in the production of distilled spirits:

$$C_6H_{12}O_6 \rightarrow 2\ CO_2 + 2\ C_2H_5OH$$

In the natural environment, CO_2 and other greenhouse gases are stored in a number of ways, under variable naturally-occurring conditions. They are stored in ocean water at various depths in ocean sediments. Carbonic acid is created by the increased levels of CO_2 in ocean water:

$$CO_4 + H_2O \rightarrow H_2CO_3$$

Carbonic acid is known to cause changes in the pH (disambiguation) levels of the oceans, stressing shellfish and coral reefs. CO_2 can also be stored in soils and vegetation. CO_2 is processed and used by natural vegetation, life forms and fauna. Carbon released into the atmosphere can be stored for long periods of time—long enough to increase atmospheric concentrations and temperatures. Figure 3-1 describes the natural carbon cycle.

The natural carbon cycle can easily be disrupted by man's activities. Electrical power generation facilities and motor vehicles account for a

Figure 3-1. The Carbon Cycle

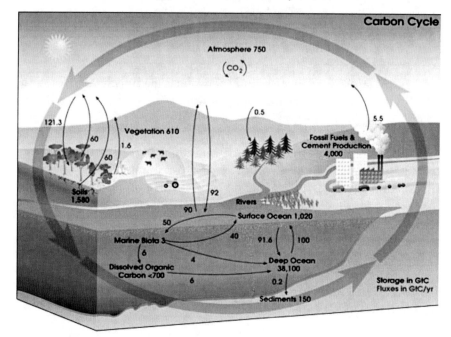

Source: NASA Earth Observatory, http://earthobservatory.nasa.gov/Library/Carbon-Cycle/carbon_cycle4.html, accessed 21 September 2007.

significant portion of sulfur oxides, carbon and nitrogen emissions that are added to the atmosphere. At some point the unnatural addition of these chemicals to the atmosphere exceeds the ability of ecosystems to absorb them without damaging results. In regard to atmospheric levels of CO_2 the following is indisputable:

- An estimated two-thirds of all past carbon dioxide emissions have been generated by the U.S. and Western European countries.[6]
- The amount of CO_2 held in the atmosphere has risen from 280 ppm, prior to the industrial revolution, to 385 ppm today and continues to increase.
- Molecules of CO_2 once suspended can remain in the atmosphere for up to 200 years.
- Rather than decreasing CO_2 emissions, the world's economies are substantially increasing them.

CO_2 emissions and overall resource use are highly correlated (Holland, Lee and McNeill 2000:87). This means that increases in energy resource consumption, especially the combustion of coal and oil, are directly related to increases in CO_2 emissions.

Extraction of carbon-based fuels can be a source of carbon emissions. To meet the growing demand for oil, some energy companies are getting very creative—extracting oil from increasingly inhospitable environments. Yet, all types of oil extraction have environmental impacts. In the Alberta province of Canada lie the Athabasca tar sands—the world's largest proven oil reserve—easily recoverable just below the rich boreal forests and peat bogs. Though much of the area is presently unreachable, Alberta's molasses-like oil reserves are 6.5 times larger than those of Saudi Arabia.[7] In Alberta, the town of Fort McMurry (sometimes nicknamed McMoney) has a wild west, boomtown atmosphere with its population growing from 32,000 to 65,000 in the last ten years.[8] While it can cost $25 to $30 per barrel to process the crude oil from tar sands, much higher oil prices make recovery economically feasible.

However, to access the oil-bearing tar sands, the land must be laid bare and energy and water resources are required. The extraction process requires heating the oil, which uses the equivalent of one barrel of natural gas for each two barrels of oil recovered. The extraction of a single barrel of crude requires two barrels of water. To accomplish this, water is drawn from the Athabasca River at a rate that would adequately supply a city with a population of one million.[9] After being used in the extraction process, much of the water is so contaminated that it cannot be returned to the river and is left in tailing ponds, which are now so large they are visible from space.[10]

Mountains of yellow sulfur, a byproduct of the oil purification process, are left abandoned from the open pit excavations that are terraforming the countryside into moon-like landscapes. Plans to extract the oil over then next ten years will leave a scar the size of Florida on the terrain of Alberta. Finally, the carbon released into the atmosphere from these operations will soon make extracting oil from tar sands Canada's largest contributor to global warming—with a carbon footprint almost equal to that of Denmark. Oil extraction processes will negate any hope of Canada meeting its commitments to the Kyoto Protocol.[11]

Yet another example is provided by China, a country with gargantuan environmental difficulties. Coal is the primary source of fuel for electric generation in China and pollution controls are lacking. In China's race to develop its economy, it has polluted its air and water to an unimaginable scale. Of the 20 major cities in the world with the most polluted air, 12 are in China.[12] Due to a number of causes, glaciers that provide melt water for China's major rivers are retreating while its deserts advance and destroy grasslands.

It seems our intent is to increase greenhouse gas emissions rather than reduce them. World emissions of CO_2 totaled 26 billion metric tons in 2004, and the International Atomic Energy Agency (IAEA) expects emissions to grow to 40 billion metric tons by 2030.[13] Given such projected increases, stabilization of CO_2e will be costly, time-consuming and potentially impossible. According to a recent report by the IPCC, full stabilization of atmospheric greenhouse gas emissions by the year 2030 would cost an extra $100 per ton of emissions.[14] It may cost even more to implement policies and programs that hope to substantially reduce them beyond a target of stabilization.

CARBON-INDUCED ENVIRONMENTAL CONCERNS

The warming of the atmosphere has become a serious environmental concern. Carbon dioxide emissions are a primary contributor to climate change. According to the U.S. Environmental Protection Agency, the ten warmest years of the 20th century all took place in its last 15 years, with 1998 being the warmest on record. The three years with the warmest recorded surface temperatures all occurred during the period from 1998 through 2003, with 2003 being the third warmest.[15] More recently, 2006 was the sixth warmest year worldwide; it was also the warmest year recorded in the U.S.[16] Summer temperatures (defined by climatologists as the months of June, July and August) have for the last 30 years exceeded the normal average worldwide.[17] However, there is evidence suggesting that 6,000 years ago, North American temperatures were actually 0.6°C (1.1°F) warmer than they are today.[18] Scientific studies regarding the earth's temperature anomalies over the last 1,000 years offer divergent results. Still, when past temperatures are reconstructed, as in Figure 3.2, the trends of the most recent 200 years are striking. There is hope that we have time to implement climate stabilization remedies, though time is running out. As it does, our mitigation options become more costly and less effective.

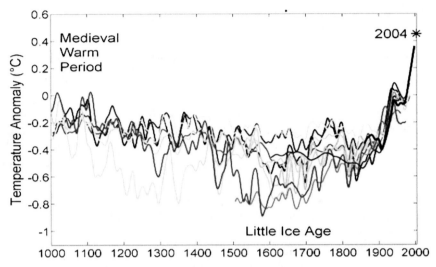

Source: Rhode, R., *Global Warming Art*. http://www.globalwarmingart.com/wiki/Image:1000_Year_Temperature_Comparison_png, accessed 2 February 2008.

Figure 3-2. Reconstructed Temperatures

The Earth's temperature has risen by 0.6°C (1.1°F) since 1900, causing an estimated 20 cm (7.9 inch) rise in sea levels (Colonbo 1992:3-14). If no effective mitigation actions are taken, carbon dioxide levels are projected to double and raise global temperatures by an additional 1°C (1.8°F) by 2025 and 1.8° to 3.5°C (3° to 6.3°F) by 2100 (Chourci 1993:51; *The Courier-Journal* 2002:A4; Union of Concerned Scientists 1999:4).

The effects of the warming are unevenly distributed. They are being felt in the northeastern U.S., where the average annual temperature has risen 1°C (1.8°F) from 1899 to 2000 and winter temperatures have risen by 2.2°C (4.0°F) from 1970 to 2000. This warming creates local climate changes. The shore along the Connecticut coast has warmed 1.7°C (3.0°F) in the past century.[19] If the warming continues, by 2100 Connecticut's climate will be similar to that of present day Virginia. Like many U.S. states, energy use in Connecticut has tripled since 1949 and CO_2 levels have increased by 17% since 1970.[20] Today, Connecticut's gaseous output equals that of Chile or Venezuela, countries that are 12 times more populous.[21] In areas of Kentucky, average daily high temperatures have increased 1.1°C (2°F) since 1947 while the average lows are 2.0°C (3.5°F) higher.[22] There, nights are getting warmer, summers longer and the fall is starting earlier in the year.

The oceans naturally absorb tremendous amounts of carbon dioxide. Increasing CO_2 levels in the world's oceans changes their chemistry. The world's oceans are now over 30% more "acidic than they were before the Industrial Revolution."[23] A recent study based on over 25,000 measurements of ocean pH has concluded that acidity due to carbon dioxide has increased more than 10 times faster than predicted by previous climate change models and other related studies.[24] Increased acidification of the oceans due to higher CO_2 levels will likely "wreak havoc" on shellfish and coral colonies, causing disruption to food chains (Vergano and O'Driscoll 2007:D2). Since natural systems are often unpredictable and difficult to model, it is also possible that at some point, the ability of ocean's to absorb CO_2 might begin to level off and possibly decline.

Changes in the climate impact life that relies on glaciers and the formation of sea ice while also increasing ocean levels. In 2008, sea ice in the Arctic melted to its second lowest level on record and scientists predict that the Arctic could be ice free in summers within ten years.[25] The Greenland and Antarctic ice sheets lose roughly 48 cubic miles of ice annually, roughly twice the amount of ice in the Alps.[26] Data collected from satellites and automated weather stations in the Antarctic (compiled by scientists in 60 countries) indicate that the melting is widespread in the Antarctic Peninsula and western regions of the continent.[27] Ice shelves that hold back glaciers are weakening. By the end of the century, sea levels due to Antarctic accelerated glacial melting may rise by a meter or more—an increase that exceeds previous predictions.[28]

Today, ocean acidification is damaging reef structures throughout the world. Coral bleaching can occur when water temperatures increase by 3°C (5.4°F) for a period of only two weeks. Coral reefs—those rain forests of the seas—are particularly susceptible to damage from higher water temperatures. Since the 1980s, an estimated one-fifth of the world's coral reefs have been eliminated—most likely caused by a combination of factors including warming conditions.[29] Worldwide, 70% of the deep-sea corals are likely to be further damaged by increased acidification and temperatures.[30]

Recent studies indicate that warming conditions have directly contributed to the extinction of at least 70 animal species, and affected approximately 59% of Earth's wild species (Vergano and O'Driscoll

2007:D2). For example, climate change could ultimately reduce the number of species in the Brazilian rain forests by as much as 30% (Watson 2007:A7).

DEFORESTATION

Climate change creates conditions that make improving forest management practices more difficult. Warming increases the potential for droughts and forest fires, which reduce the ability of forests to process and store carbon dioxide. Roughly 20% of the world's global warming gases are caused by deforestation. While the total impact is unknown, it has been estimated that these actions cause more heat-trapping CO_2 to be exhausted into the atmosphere than the combined emissions from "all the world's trains, trucks and automobiles."[31] A forest area the size of 300 soccer fields is cleared every hour in Indonesia and illegal logging costs the government $3 billion in lost revenues annually.[32] In Africa, 130,000 km² (50,000 sq. miles) of tropical forests are being cleared annually for farming.[33] In the Amazon and other South American countries forests are burned for cattle, sugar cane and soybean production; in Indonesia island forests are cleared for palm oil production.[34] In the month of September 2008 alone, it was estimated that 777 km² (300 square miles) of rain forests were illegally burned to clear land for farming in Brazil—three times the amount lost in September 2007. When considered on a global sale, 24 hectares (60 acres) of tropical forests disappear every minute and dying forests cause climate change to accelerate.[35]

Once such natural ecosystems are destroyed they are very difficult, if not impossible, to recreate. Maintaining and restoring natural forest ecosystems is one way to sequester carbon. There are aggregators of carbon credits that attempt just this. The Forest Opportunities Initiative began in 2006. This program uses forest management and land set-aside initiatives to increase the inventory of land available for biomass sequestration. In its first year, over 4,850 hectares (12,000 acres) of forests were placed into management and certified by forest management practitioners. Depending on the type of forest, the ability of forests to store carbon varies from 4 to 9 metric tons per hectare (1.6 to 3.6 metric tons per acre).

Pasture and cropland capture roughly 7.4 metric tons of CO_2 per hectare (3 tons per acre) annually. Afforestation initiatives are feasible.

For example, the U.S. with 176 million hectares (435 million acres) of cropland has the ability to reforest approximately 13 million hectares (32 million acres)—an initiative with the potential to offset 80 megatons of carbon emissions annually.[36] These initiatives create new opportunities for landowners to obtain income from their land. Revenue from the sale of carbon credits supplements the landholder's income, creating incentives to maintain forests rather than harvesting the timber and other resources found on the land. Income from the carbon storage potential creates supplemental incentives to leave productive forests in place.

THE KYOTO PROTOCOL

The United Nation's (UN) Climate Change Summit in Kyoto, Japan in 1997 was a follow-up to the Framework Convention on Climate Change at the Earth Summit in Rio in 1992. The Kyoto Summit brought 160 nations together in an effort to reduce carbon emissions and devise ways to mitigate climate change. The countries involved offered varying perspectives on the issues, perspectives based on their potential costs, benefits, resources, and differences in scientific knowledge. At Kyoto, the success of the negotiations depended on the individual nations' willingness to comply with the treaty. This proved to be a difficult task since each nation proposed different methodologies to reduce emissions.

The agreement reached is known as the Kyoto Protocol. The Protocol required commitments from the 39 industrialized countries to reduce emissions of greenhouse gases to a level 5% below their 1990 emission levels by 2012 (Kraft 2004:274). As one example, Canada agreed to reduce emissions by 6% below 1990 levels. Included in the Kyoto Protocol as a means of reducing greenhouse gases, the Clean Development Mechanism (CDM) provides guidelines for Joint Implementation (JI) and International Emissions Trading (IET). Certified Emission Reductions (CERs) are generated by the CDM under Article 12 of the Protocol.

The guidelines are flexible and allow developed countries to meet their reduction commitments through domestic efforts or through joint projects with other countries. The CDM and IET provide opportunities to develop a global carbon trading market through which companies in the industrialized world "earn credit not for reducing their own emissions but for investing in energy efficiency improvements in the developing world."[37] The types of credits that are defined in the Protocol include:[38]

- Assigned amount unit (AAU)—an emission-trading unit issued by a party to the Protocol (under the Marrakech Accords) equal to one metric ton of CO_2 equivalent.
- Certified emissions reduction (CER)—examples include the Prototype carbon fund and the Dutch Certified Emission Reduction Unit Procurement Tender (CERUPT) program.
- Emission reduction unit (ERU)—a reduction in greenhouse gas emissions that represents one metric ton of CO_2 equivalence.
- Removal unit (RMU)—an emissions unit issued by a party to the Protocol on the basis of land use changes and forestry.

Kyoto-compliant projects using carbon offsets are becoming popular. Hundreds of projects, mostly in Brazil, China, and India, when fully implemented, will reduce CO_2 emissions by the equivalent of 115 million tons annually.[39] Rajesh K. Sethi, Secretary of India's CDM Authority, calls the program "one of the most successful ways we have found to reduce greenhouse gases."[40] A registry system is used to track and record CER transfers. Often, it is less expensive for developed countries to implement carbon-reduction projects beyond their borders rather than within their own countries. For this reason the Kyoto Protocol has its critics. Canadian Prime Minister Stephen Harper has referred to Kyoto as a "socialist scheme designed to suck money out of rich countries."[41]

U.S. INITIATIVES AND REACTIONS
TO THE KYOTO PROTOCOL

The U.S. emissions reduction target proposed by Kyoto for 2012 was a 7% reduction below 1990 levels (Kraft 2004:274). Meeting this goal would not have seemed difficult if energy consumption and carbon emissions had been stabilized at 1990 levels. Today, however, such a goal seems daunting.

To meet this goal, the U.S. would need to increase automobile fuel economy standards by 10% and reduce coal use in power plants by roughly 10%. In 2006, U.S. manufactured vehicles had an average fuel efficiency of 10.8 km/l (25.4 mpg) while those manufactured by Toyota in North America averaged 14.7 km/l (34.7 mpg)—36% greater fuel economy.[42] Since most coal is used to generate electricity, a CO_2 reduction of the magnitude required might be achieved by improving energy efficiencies and instituting demand reduction programs, which would upgrade residential and commercial building codes. In addition, utilities in the U.S. would be required to satisfy new electrical demand by using alternative fuels.

While the 15 members of the European Union (EU) (out of a total of 169 countries worldwide) have ratified the Kyoto Protocol, the United States has not.[43] President Bill Clinton was a signatory to the Kyoto Protocol, but President George W. Bush withdrew from it, claiming that it was fatally flawed because of its adverse economic impact and provisions that excluded developing countries from complying with its standards. Regardless, G. W. Bush's campaign promise in 2000 assured the nation that "CO_2 would be regulated as a greenhouse gas."[44] This changed quickly and this promise was unfulfilled during his presidency. Quin Shea of the Edison Electric Institute stated bluntly in 2001: "Let me put it to you in political terms... the President needs a fig leaf. He is dismantling Kyoto."[45] While Kyoto was not dismantled, the U.S. functionally abstained. As late as 2008, the U.S. remained "the only major industrial country to have rejected the Kyoto treaty and its obligatory targets."[46] As it turned out, the lack of improved fuel efficiency standards did far more damage to the U.S. economy when oil prices topped $140 per barrel in 2008 than attempting to comply with the Kyoto Treaty.[47]

In January of 2007, the president issued Executive Order 13423, mandating in Section 2 that Federal agencies in the U.S. reduce energy use by 3% annually, or a total of 30% through 2015.[48] The order required fleet consumption of petroleum-based fuels to be reduced by 2% annually over the period. Furthermore, non-petroleum-based fuels would act as a substitute and be increased 10% annually.[49] Section 3(a) of the order requires agencies to implement "sustainable practices for energy efficiency, greenhouse gas emissions avoidance or reduction, and petroleum products use reduction."[50] Though the Executive Order was mandated, the necessary funding for its implementation was neither identified nor provided. To date, there is no documentation addressing whether these goals were achieved in 2007 or 2008.

While acknowledging the existence of climate change, the G.W. Bush administration continued to reject mandatory limits on greenhouse gases.[51] To date, the U.S. government has not yet offered a workable alternative to the Kyoto Protocol. CO_2 remains an unregulated greenhouse gas in the U.S. For its part, the G.W. Bush administration favored voluntary reductions in greenhouse gases and further research to study the issues of global warming and climate change.[52] His administration claimed that adherence to the Protocol would cost the U.S. as much as $400 billion to implement and that 4.9 million jobs would be eliminated. The perception was that the economic payoffs for the U.S. as a 'free rider' were simply greater than supporting the Kyoto Protocol and the coalition against climate change (Zhang 2004:443).

In retrospect, estimates of economic and employment losses of the magnitude suggested by the G.W. Bush administration appear to be patently unsupportable and politically motivated. The projected costs were more than half of the total U.S. expenditures for energy in the year 2000. Predicted job losses are a multiple of those estimated by the same U.S. administration to have been lost during the 2001 recession. Such inordinate estimates of job losses were unprecedented and largely considered only for their short-term impacts on incumbent industries. Interestingly, the employment increases created by implementing sustainable development policies, such as marketing environmental technologies, were overlooked in the analysis. For example, U.S. companies exported $2.3 billion worth of environmental goods to China in 2006, three times the amount in 2003.[53] Transition Energy, a General Electric (GE) subsidiary, increased exports almost 150% during this period. GE also opened a factory in Shenyang that produces 150 to 300 1.5-megawatt (MW) wind turbine generators annually and has since captured 13% of the market in China.[54]

Also omitted from the analysis were estimates of the employment that would be generated by developing renewable energy technologies, energy services, environmental mitigation efforts, and the exportation of equipment and engineering expertise that would offset any anticipated job losses. By 2007, the U.S. wind energy industry had a total installed capacity of 5,244 MW, an increase of 45% over 2006, and was estimated to employ over 35,000 people.[55] The net gains to the U.S. economy resulting from energy efficiency improvements at the macroeconomic scale were also overlooked. Finally, the analysis failed to consider the national security implications of energy supplies, the impact of imported energy on trade deficits, the economic costs of currency devaluations, and the financial risks associated with escalating imported energy costs.

Meanwhile, the EU has committed to binding reductions of 20% in greenhouse gases by the year 2020.[56] This leadership is paying big dividends. The experience of Britain provides a real-world indicator of the impact on employment. Rather than causing economic harm and significant employment losses, complying with the Kyoto Protocol has had the opposite effect. According to former Prime Minister Tony Blair, "there are great commercial opportunities to be had" (Breslau 2007:51). More than 500,000 "clean tech" jobs have been created since Britain began complying with the Kyoto Treaty (Breslau 2007:51). Many of these jobs are technology-oriented and have resulted in higher-than-average pay scales.

TXU, Texas' largest utility, provides an example of this distinctive shift in direction and the opportunities it can offer. As TXU faced concerns by environmentalists regarding its plans to construct 11 coal-fired, electrical-generating plants, a leveraged buy-out was arranged. After the deal was completed, plans for eight of the coal-fired plants were withdrawn. There were concerns by the utility's new owners that emission caps would eventually create substantial penalties for carbon-based generating facilities. As an alternative, TXU instead decided to invest $400 million on renewable energy projects and other conversion initiatives (Davidson 2007).

Despite this dramatic change in direction for a single company, approximately 150 coal-fired power plants are planned for development in the U.S. However, while the U.S. Congress is seeking workable ways to reduce greenhouse gases, a program established in 1935 actually encourages rural electrical cooperatives to construct coal-fired plants. Many of the service territories of the co-ops, which were once rural areas, are now actually in the suburban areas of large cities. The program from the Agriculture Department's Rural Utilities Service provides subsidized, low-interest, federal loans to develop power plants. Rural electric co-ops plan to use the program to invest $35 billion in conventional coal plants over the next 10 years.[57] Rural co-op electrical production is mostly derived from coal, and electrical demand in rural areas is increasing at twice the national rate.[58] Since these coal-fired plants generate greenhouse gases, the Agriculture Department's program effectively provides financing to "offset all state and federal efforts to reduce U.S. greenhouse-gas emissions" over the next 10 years.[59] There are no comparable federal programs that subsidize long-term financing of electrical energy production from alternative energy sources.

Environmental problems are caused by the extraction of coal. Mining coal is a major industry and supports local economies worldwide. There are many ways to mine coal including deep mining and strip mining. The most controversial extraction technique is called "mountaintop removal," a method whereby the tops of hills are blasted away to expose the coal seams. Any debris from the removal is pushed down the hillsides, filling valleys and covering streams. It is used in states such as Kentucky and West Virginia—and natural ecosystems are destroyed in the process. However, the environmental destruction that accompanies coal mining generally devalues the surrounding area. Once the coal has been mined out (typically 10 to 15 years) the communities are left with unemployment and lower-quality ecosystems.

Yet no amount of environmental remediation will ever restore the quality of the ecosystems that existed prior to mining. According to the U.S. Environmental Protection Agency, mountaintop removal accounted for less than 5% of U.S. coal production in 2001, yet hundreds of thousands of acres of forests have been destroyed and hundreds of mountains have been leveled.[60] The destruction of these forests reduces nature's ability to capture and store CO_2. In addition, coal bed methane gas is released into the atmosphere during this mining, contributing to greenhouse gas emissions.

Oil drilling also impacts the environment. Drilling for fossil fuels varies with location and drilling practices, but may include any of the following: disturbances to wildlife and vegetation, the release of drilling fluids, contamination of groundwater and surface water, seismic disruptions, and gaseous emissions. Transporting these fuels creates opportunities for catastrophic accidents and terrorist acts. Oil spills have caused massive damage to the environment all over the world, whether accidental or intentional. Incidents include the 400,000 barrels of oil spilled from a major pipeline in the Amazon region of South America, the 38,000 barrels spilled in Nigeria, and the 28,500 barrels spilled in Columbia.[61] The flaring of gas wells also poses serious local environmental problems. Methane emissions from oil and gas operations, along with coal bed methane and agriculturally produced methane, make up an estimated 9% of the U.S.'s GHG emissions.[62]

There are alternatives for the U.S., such as nuclear power, that may have a greater role to play. Generating electricity from nuclear energy production generates no greenhouse gases except in fuel extraction and processing. According to Jerry Paul at the Howard Baker Center for Public Policy, "It is all about climate change and emissions. It is about economics and the recognition that nuclear power has the lowest operating cost of any form of (electrical) base load generation."[63] Tennessee Valley Authority's (TVA) Brown's Ferry Unit #1 nuclear plant came on line in 2007, after $1.8 billion in renovations (the plant was mothballed after a fire in 1985). It is the first "new" reactor since 1996 in the U.S. There are problems with the idea of expanding nuclear power. These include the high costs of construction, the long lead time for development (10 years of more), the lack of waste disposal facilities and the national security implications of centralized power generation. Regardless, the U.S. Department of Energy (USDOE) estimates that an additional 50 nuclear power plants will be required by 2030.[64]

While technologies are available, the goal of reducing greenhouse gases continues to be elusive in the U.S. and elsewhere. The bottom line is that the actual results of the Kyoto Protocol have been mixed—and in some cases ineffective. Most nations have yet to baseline their production of global climate change gases. Data from a UN report indicated that from 2000 to 2004, 34 of the 41 countries it monitors (excluding most developing countries) had actually increased greenhouse gas emissions.[65] This sobering assessment provides evidence of how challenging reducing greenhouse gas emissions can be in real-world applications and how difficult it is for countries to meet greenhouse gas reduction goals.

THE MONTREAL PROTOCOL

One of the biggest obstacles in managing environmental problems is that many environmental issues are seen as international in scope rather than local. Existing institutions and national bureaucratic structures often cannot handle and were not created to deal with these problems effectively (Cusimano 2000:7). Supra-national governmental organizations such as the UN and the EU have proved more effective. Without the institutional capacity in place to handle problems on an international scale, inertia and inaction are often the result. International action to prevent additional damage to the ozone layer required an international treaty. The effect of ozone-depleting gases, which contributed to the "ozone hole" in the atmosphere, provides an example of a success story where international cooperation has yielded positive results.

The ozone layer in the upper atmosphere protects life from the harmful effects of ultraviolet radiation. Processes that deplete the ozone layer magnify these effects. It was U.S. and British scientists who first determined that there was an opening in the ozone layer over the Antarctic. While Sweden became the first nation to ban aerosol sprays that were thought to damage the ozone layer, the Antarctic ozone hole was actually not discovered until 1985.[66] Human effects of exposure to ultraviolet radiation include severe sunburn and epidermal cancers.

The Montreal Protocol on Substances that Deplete the Ozone Layer, known as the Montreal Protocol, was an important international environmental agreement. It was signed by 160 countries in order to control the production of ozone-depleting compounds. By establishing a timetable to phase out the production and use of ozone- depleting substances, it was hoped that further damage would be avoided, and that the ozone layer could be repaired. However, after reaching a scientific consensus the production of ozone-depleting substances was not quickly phased-out, but rather evolved over decades.

Chlorine- and bromine-containing gases, such a chlorofluorocarbons (CFCs) and halon, are known to destroy ozone molecules in the troposphere. According to the terms of the Montreal Protocol, the manufacture of halogenated CFCs (Group I substances) would be reduced to just 50 percent of 1986 levels by 1999. In addition, manufacture of halon gases (Group II substances) were frozen at 1986 levels by 1992.

The Montreal Protocol incorporated a procedure for reviewing and refining the agreement if scientific information changed (Andrews 1999:328; Elliott 1998:56). Since the treaty was signed by the participating nations in 1987, and went into effect January 1, 1989, there have been five different revisions of the Protocol to update and strengthen the phase-out timetables of ozone-depleting gases, including a complete phase-out of CFCs, halon gases, and carbon tetrachloride by 2000 (Elliott 1998:58).

The Montreal Protocol is widely viewed as one of the most successful environmental agreements ever adopted (Kraft 2004:270). Regardless, the Protocol has numerous flaws that have undermined its success. One clause effectively postponed implementation by including a 10-year grace period for compliance by developing counties, as long as their consumption of ozone-depleting substances did not exceed 0.3 kg (.66 lbs.) per capita. Many of the developing countries chose not to participate in the initial negotiations of the protocol, although some eventually ratified the treaty (Elliott 1998:56-57). While there are voluntary reporting provisions at national levels, there are no enforcement or verification procedures through international oversight (Elliott 1998:57). Since the initial protocol was approved, subsequent amendments adopted after the protocol's approval have yet to be ratified by many countries.

As a result of the Montreal Protocol, the ozone layer is improving. Measurements of ozone-depleting gases indicate that they peaked in 1994 and have since declined. By 2006, ozone levels had improved by 22% in the mid-latitudes. There was also a 12% reduction in conditions that had initially resulted in the development of the ozone hole.[67] If present trends continue, the recovery of the mid-latitude ozone layer will occur about 2045-2055, while the full the recovery of the ozone layer is not expected until 2075-2080.[68]

While this is encouraging, the recovery of the ozone layer may be in jeopardy due to increasing levels of greenhouse gases in the atmosphere. We know that the problems associated with greenhouse gas emissions such as global climate change may not be resolved during this century, regardless of any presently available mitigation actions that might be undertaken. By warming the atmosphere, greenhouse gases such as methane and CO_2 will cause further damage to atmospheric ozone molecules. This is unsettling to atmospheric scientists. According to a National Aeronautics and Space Administration (NASA) study, "Climate change from greenhouse gases can also affect ozone by heating the lower stratosphere where most of the ozone exists. When the lower stratosphere heats, chemical reactions speed up and ozone gets depleted."[69] While there is uncertainty, computer models indicate that an increase in greenhouse gas emissions may impede and delay the recovery of the ozone layer.

DISEASES

Climate change may be causing the proliferation of certain diseases, especially in remote locations. Growing atmospheric CO_2 concentrations have been documented, and are attributed with increases in certain diseases. Dr. Ivo Mueller, a scientist at the Papua New Guinea Institute of Medical Research, noted, "malaria epidemics in the highlands are now basically happening every year."[70] Many in the country live at altitudes above 1,500 meters (5,000 ft.)—where malaria in the past has been a rare occurrence. Warming temperatures now allow mosquito populations that carry the parasite causing malaria to invade locations at higher elevations where it was previously too cold for them to flourish. Cases of malaria in the Western Highlands have increased from 638 in 2000 to 4,986 in 2003.[71] In many poorer countries, the health-care infrastructure is not in place to meet the demands for treatment now needed.

SUMMARY

We now know that the effects of climate changes are far-reaching and that carbon is the culprit. Global warming causes disruption to natural ecosystems and increases the potential for droughts and forest fires, which reduce the ability of forests to process and store carbon dioxide. Once such natural ecosystems are destroyed they are very difficult, if not impossible, to recreate. Increasing temperatures can also cause diseases to become more pervasive. The economic implications must be considered. According to Daniel Lashoff, director of the Climate Center at the National Resources Council, "This debate is going to be mostly about costs... but we want to make sure... we don't forget that the costs of inaction would be much higher than the cost of the emission reductions."[72] Failure to take action to reduce CO_2 emissions exacerbates environmental damage.

Hydrocarbon-based energy consumption is increasing. Carbon dioxide is released into the atmosphere by the combustion of carbon-containing fuels and petroleum-based distillates. Atmospheric carbon and GHG levels that drive global climate change have the potential to negate our boldest efforts towards mitigation. How we use energy is key to our mitigation efforts. Renewable energy sources can be categorized as *sustainable*, while most nonrenewable energy sources are *potentially unsustainable*, and likely exhaustible. Refocusing our efforts to develop renewable energy resources is one solution. There is hope that we have time to implement climate stabilization remedies.

The Kyoto Summit brought 160 nations together in an effort to reduce carbon emissions and devise ways to mitigate global warming. The agreement reached is known as the Kyoto Protocol. It required commitments from the 39 industrialized countries to reduce emissions of greenhouse gases to a level 5% below their 1990 emission levels by 2012 and provided guidelines for implementation. While technologies are available, the goal of reducing greenhouse gases continues to be elusive. The actual results of the Kyoto Protocol have been mixed and in some cases ineffective.

The Montreal Protocol on Substances that Deplete the Ozone Layer was an important international environmental agreement. It attempts to control the production of ozone-depleting compounds. By establishing a timetable to phase out the production and use of ozone-depleting substances, it was hoped that further damage would be avoided, and that the ozone layer could be repaired. Since the ozone layer is beginning to recover, the Montreal Protocol is widely considered to be successful. It offers hope that with international mitigation efforts, we can control damaging substances such as greenhouse gases that are being released into the atmosphere. However, by raising temperatures, increasing CO_2 levels in the atmosphere may hinder our success in reducing atmospheric ozone levels.

Endnotes

1. Duke Energy (2007, September 27). *Eight utilities seek to increase energy efficiency investment $500 million annually.* http://www.duke-energy.com/news/releases/2007092701.asp, accessed 5 October 2007.
2. Schloerer, J. (1996, October). *Why does atmospheric CO_2 rise?* http://www.radix.net/~bobg/faqs/scq.CO2rise.html, accessed 20 September 2007.
3. U.S. Geological Survey (2006, January 10). *Volcanic gases and their effects.* http://volcanoes.usgs.gov/Hazards/What/VolGas/volgas.html, accessed 20 September 2007.
4. *Ibid.*
5. Pierantozz1, R. (2001). Carbon Dioxide. *Kirk-Othmer Encyclopedia of Chemical Technology.* Wiley.
6. Revkin, A.C. (2007, April 1). Poor nations bear brunt of warming. *New York Times.*
7. Edemariam, A. (2007, December 9). Dark side of the boom. *Postmagazine.* Hong Kong. p. 28-32.
8. *Ibid.*
9. *Ibid.*
10. *Ibid.*
11. *Ibid.*
12. Lynch, D. (2007, September 18). China's 'grappling with one helluva problem', *USA Today.* p. 1B.
13. Kaufman, M. (2007, February 7). *Global warming and hot air.* www.washingtonpost.com/wp-dyn/content/article/2007/AR200702060126.html, accessed 6 May 2007.
14. Kaufman, M. (2007, May 5). Scientists rate costs of reducing global warming. *The Courier-Journal.* p. A7.
15. Records have been kept by the World Meteorological Organization since 1861. The average temperature of Earth's surface has varied between 13.8 and 14.6 degrees Celsius (56.8 and 58.3 degrees Fahrenheit) during the period from 1950 to 1999. In the year 1999, the average global temperature was approximately 14.4 degrees Celsius (57.9 degrees Fahrenheit). See http://wiki.answers.com/Q/What_is_the_average_temperature_of_earth's_surface, accessed 11 September 2008. December makes '06 warmest year in U.S. (2007, January 10). *The Courier- Journal.*
16. Data for the U.S. was accumulated from 1,200 temperature recoding stations across the country.
17. No relief in sight for summer heat trend. (2008, September 17) *USA Today.* p. 4D.
18. Holahan, D. (2007, November). Just warming up. *Connecticut Magazine.* p. 77.
19. *Ibid.,* p. 57.
20. *Ibid.,* p. 76.
21. *Ibid.*
22. Bruggers, J. (2008, October 20). Our changing climate—Unprecedented study

exams 60 years of weather. *The Courier-Journal*. p. A4.

23. Lovejoy, T. (2007, April 16). The ocean's food chain is at risk. *Newsweek*. p. 80.
24. Hogan, D. (2008, December 8). A sea of change. *The Courier- Journal*. p. E3. The study referenced "Dynamic patterns and ecological impacts of declining ocean pH in high-resolution multi-year dataset" was by J. Timothy Wootton, published 2 December 2008.
25. No relief in sight for summer heat trend. (2008, 17 September). *USA Today*. p. 4D.
26. Eilperin, J. (2009, January 5). Faster climate change. *The Courier- Journal*. p. D3.
27. Engeler, E. (2009, February 26). Antarctic glaciers melting faster than thought, report says. *The Courier-Journal*. p. A7.
28. *Ibid.*
29. Philips, M. (2007, July 9). The fading forests of the sea. *Newsweek*. p. 50.
30. UNESCO (2007, April 10). *Climate change threatens UNESCO world heritage sites.* whe.unesco.org/pg_friendly_print.cfm?id=319&cid=82&, accessed 2 July 2007.
31. Harris, E. (2008, February 3). *Felling of rain forests imperils climate.* http://www.commercialappeal.com/news/2008/feb/03/felling-of-rain-forests-imperils-global-climate/, accessed 2 March 2008.
32. Hot flat and crowded (2008, October). *Sky*, p. 42-49.
33. Global tree-felling imperils climate. (2008, February 3). *The Courier-Journal*. p. A20.
34. *Ibid.*
35. *Ibid.*
36. McKinsey & Company (2007, December). *Reducing U.S. greenhouse gas emissions: how much at what cost?* U.S. greenhouse gas abatement mapping initiative, http://www.mckinsey.com/clientservice/ccsi/greenhousegas.asp, accessed 3 February 2008. p. 55.
37. Robinson. S. (2007, April 14). The global warming survival guide - Trade carbon for capital. *TIME*. www.time.com/specials/2007/environment/article/0,28804,1602354_163074_1603643,00.html, accessed 20 May 2007.
38. Climate Change Projects Office (2005, April). *Carbon prices.* United Kingdom Department of Trade and Industry.
39. *Ibid.*
40. *Ibid.*
41. CBC News (2007, January 30). *Harper's letter dismisses Kyoto as 'socialist scheme.'* http://www.cbc.ca/canada/story/2007/01/30/harper-kyoto.html, accessed 3 June 2007.
42. On 21 June 2007, the U.S. Senate approved legislation to increase fleet fuel economy standards to 35 miles per gallon by 2020 for cars, sport utility vehicles and some trucks.
43. The actual agreement of the Kyoto Protocol Conference of Parties (COP) in December 1997 was COP-3 which created a global target reduction in six primary greenhouse gases (including CO_2) of 5.2 percent from 2008 to 2012 (Chasek 2000:77-78).
44. Gore, A. (2007). *The assault on reason.* New York: Penguin Press. p. 194.
45. *Ibid.*
46. Hanley, C. (2007, December 9). U.S. 'not ready' to commit to emission cuts at meeting. *The Courier-Journal*. p. A13.
47. By March 1, 2009, the price of benchmark crude oil had fallen to $44 per barrel.
48. Office of the Presidential Press Secretary (2007, 24 January). *Executive Order: Strengthening federal environmental, energy, and transportation management.* http://www.whitehouse.gov/news/releases/2007/01/print/20070124-2.html, accessed 6 July 2007. Agency heads in intelligence, military, and law enforcement may exempt their departments.

49. *Ibid.*, applies to agencies with 20 or more vehicles.

50. *Ibid.*

51. Borenstein, S. (2007, February 6). Report: Global warming 'very likely' man-made, unstoppable for centuries. *The Courier-Journal.* p. A5.

52. Kraft, M. (2004). *Environmental policy and politics*, Third Edition. New York: Pearson/Longman. p. 51.

53. Lynch, D. (2007, September 18). Opportunity shines in hazy days of China. *USA Today.* p. 1B.

54. *Ibid.*, p. 2B.

55. American Wind Energy Association (2008, January 17). *Installed U.S. wind power capacity surged 45% in 2007.* http://www.awea.org/newsroom/releases/AWEA_Market_Release_Q4_011708.html, accessed 14 March 2008.

56. Hanley, C. (2007, December 9). U.S. 'not ready' to commit to emission cuts at meeting. *The Courier-Journal.* p. A13.

57. Mufson. S. (2007, May 14). Federal loans fuel new coal plants. *The Courier-Journal.* p. A3.

58. *Ibid.*

59. *Ibid.*

60. U.S. Environmental Protection Agency (2005, October). *Mountaintop Mining/Valley Fills in Appalachia: Final Programmatic Environmental Impact Statement.*

61. Trade and Environment Database, *Oil Production and Environmental Damage*, http://www.american.edu/TED/projects/tedcross/xoilpr15.htm#r3, American University.

62. Energy Information Administration, U.S. Department of Energy, *Greenhouse gases, climate change and energy.* http://www.eia.doe.gov/oiaf/1605/ggccebro/chapter1.html.

63. Mansfield. D. (2007, May 5). TVA sees future in nuclear power. *The Courier-Journal.* p. D2.

64. *Ibid.*

65. Samuelson. P. (2006, November 10). *Greenhouse guessing.* www.washingontonpost.com/wp-dyn/content/article/2006/11/09/AR2006110901768.html?nav=rss_opinion/columns, accessed 6 May 2007.

66. *Ozone layer.* http://en.wikipedia.org/wiki/Ozone_layer, accessed 6 June 2007.

67. Data from the National Oceanic and Atmospheric Association, *NOAA/ERSL Releases Ozone Depleting Gas Index*, accessible at www.noaa.gov/hotitems/storyDetail_org.php?sid=3843, dated 15 January 2007.

68. *Ibid.*

69. Goddard Space Flight Center (2002, June 4). *Climate change may become major player in ozone loss.* www.gsfc.nasa,gov/topstory/2002041greengas.html, accessed 6 May 2007.

70. Hanley, C. (2007, December 9). Malaria's spread on island blamed on global warming. *The Courier-Journal.* p. A13.

71. *Ibid.*

72. Hebert, H. (2008, June 2). Can we fix global warming? *The Courier-Journal.* p. A3.

Chapter 4

Governmental Carbon Reduction Programs

"The United States could reduce greenhouse gas emissions in 2030 by 3.0 to 4.5 gigatons of CO_2 emissions using tested approaches and high-potential technologies. These reductions would involve pursuing a wide array of abatement options available at marginal costs less than \$50 per ton, with the average net cost to the economy being far lower if the nation can capture sizable gains from energy efficiency. Achieving these reductions at the lowest cost to the economy, however, will require strong, coordinated economy-wide action that begins in the near future."

McKinsey & Company (December 2007)

Governments establish frameworks of incentives to both mandate and encourage changes in environmental policies. These policies are manifested in their laws, legislative practices and enforcement methodologies. In the past, much of the legislation was focused on increasing energy resources and encouraging economic development. Development of energy resources is a means of growing the state economy, especially when energy resources are abundant and easily exploitable.

However, exploiting easily recoverable energy resources, such as coal and oil, often results in adverse environmental consequences. Such consequences include pollution of the air and water. Since the 1960s and 1970s, environmental legislation has focused on mitigating the environmental impacts caused by this exploitation and use. As the impact of high atmospheric carbon concentrations became evident at the turn of the century, governments began passing legislation focusing on carbon mitigation. Indeed, government-enforced carbon mandates appear to be the new wave of the future.

There are many variations in the patchwork of unaddressed environmental problems that form new legislative agendas. The variance found within and between federal and state-sponsored initiatives exemplifies the complexity of mitigation efforts. In addition, the threat of global climate change has instigated numerous international treaties, such as those developed with the UN and EU's oversight.

INITIATIVES BY FEDERAL GOVERNMENTS

Carbon emissions are expected to continue to increase throughout the world during the next 20 years. Average annual growth rates are projected to be 1.5% in the U.S., 3.0% in India, and 3.4% in China during this period.[1] As a result of the Kyoto Protocol, governments are taking action. Actions include setting goals for the use of fuels that do not contribute to carbon emissions. In the EU, a target has been established for member countries to obtain 12% of all energy and 22% of electricity from renewable resources by 2010.[2]

One nearly carbon-free electrical generation technology that is experiencing a renaissance is nuclear energy. Governments worldwide are supporting the construction of nuclear power plants. More than 100 nuclear reactors are under construction, planned or on order, and about half are located in India, China and other developing nations.[3] Though 20% of Great Britain's electricity is now generated by nuclear power plants, most will likely be decommissioned by 2030. Consequently, the government is supporting the construction of new nuclear power plants throughout the country.[4] Other countries such as Argentina, Brazil and South Africa are also planning to expand nuclear energy production.[5] Some developing countries are planning to construct their first nuclear reactors. This revival could almost double the world's output of nuclear-generated electricity within the next 20-30 years. However, safety and radioactive waste management issues remain unresolved.

In Australia, the government's climate change strategy seeks to reduce the country's "greenhouse gas emissions, adapt to the climate change we cannot avoid and help shape a global solution."[6] The country emitted the equivalent of 576 million metric tons of CO_2 in 2006.[7] Australia's carbon pollution reduction scheme was launched in 2008 with the intent of developing a robust carbon market. This plan focuses on the nation's 1,000 largest polluters or those that emit more than 25,000 tons of carbon annually. The government has committed to reducing emissions by 60% from 2000 levels by 2050. Their national cap-and-trade program involves a four-step process:[8]

Step 1: Significant emitters of greenhouse gases will need to acquire a 'carbon pollution permit' for every ton of greenhouse gas emitted.
Step 2: The quantity of emissions produced by the companies involved is monitored and audited.
Step 3: At the end of each year, each liable firm will surrender a 'carbon pollution permit' for every ton of emissions produced during that year.
Step 4: Firms compete to purchase the number of 'carbon pollution permits' that they require. Firms that require carbon permits will need to purchase them at auction or in secondary markets.

As a result of this program, many firms will find that it is less expensive to reduce emissions than to continue purchasing permits. Carbon credits are issued using a program called the "Australian Government Friendly" scheme. A third party that has been approved by the Australian Greenhouse Office must independently verify carbon credits. Providers of carbon credits must meet rigorous standards that prevent double accounting of the carbon emissions credits issued.

The carbon reduction goals of countries develop from their special circumstances and the needs of their economies. Worldwide, 20% of the world's methane gas emissions are generated from livestock. New Zealand is a country of 4 million people with few major industries. Interestingly, the major source of greenhouse gases happens to be methane gas from livestock including 40 million sheep and 9 million cattle.[9] Charlie Peters, a New Zealander once commented that "there's no other country in the world that's so clean of chimney stacks that its animals are the biggest polluters."[10] Regardless, New Zealand has established a goal of becoming the world's first major carbon neutral country—a daunting yet notable goal.

INTERNATIONAL PUBLIC-PRIVATE PARTNERSHIPS

International public-private partnerships are being established with various goals which include standardizing best practices, iden- tifying alternative energy development projects, co-development opportunities, and technology transfer opportunities. These are often targeted collaboration efforts, each with a specifically defined set of goals. The Climate VISION program and the Climate Leaders program are examples of voluntary public-private collaborations between corporations and industry to reduce greenhouse gas emissions. The Carbon Sequestration Leadership Forum is a multinational collaborative effort with a goal of developing cost-effective methods to sequester carbon from coal, whose use remains an economically attractive energy option for many countries.[11] The International Partnership for the Hydrogen Economy organized 16 developed and developing countries to coordinate international research, development, and commercialization of hydrogen technologies.[12] For nuclear energy development, the 11-member Generation IV International Forum is working on new atomic reactor designs that hold potential for hydrogen production.[13] However, there are less complex and costly ways to produce hydrogen in commercial quantities.

The Asia-Pacific Partnership (APP) on Clean Development and Climate is a cooperative effort that includes Australia, Canada, China, India, Japan, Korea and the U.S.—countries that collectively represent more than half of the world's economic base, total population and energy use—plus 54% of global CO_2 emissions from fossil fuel consumption.[14] These countries jointly produce 39% of the world's renewable energy for electrical power generation. As of 2005, they generated 146 billion kWh from energy derived from geothermal, solar, wind and biomass wastes.[15]

The APP partnership includes governmental, industrial, commercial, and university entities. The APP was formed to address "increased energy needs and the associated issues of air pollution, energy security and climate change."[16] The objectives of the partnership include promoting economic development, reducing poverty and accelerating "the development and deployment of clean, more efficient technologies."[17] Their efforts are focused on leveraging the expertise of public and private entities in eight major energy consuming sectors: aluminum, buildings and appliances, cement manufacturing, cleaner use of fossil fuels, coal mining, power generation and transmission, renewable energy and distributed generation, and steel. This is a broad charter that will be difficult to implement.

According to U.S. Ambassador Reno L. Harnish III, the "APP is unique and funded by the U.S. government to accelerate green energy development" and provides a "structural approach to solving climate change problems."[18] The APP is attempting to further expand investment and trade in cleaner and more efficient technologies that are directed toward:[19]

- Speeding deployment of existing cleaner technologies
- Developing long-term transformational technologies
- Promoting improved policy environments
- Creating a forum for the exchange of clean energy best practices

The partnership has over 100 projects and activities under development. To support the APP, the U.S. Department of State has co-funded 12 clean energy technology development projects in India with a total funding of $6 million that were selected using a competitive solicitation process, plus an additional $11 million in grants for similar projects in both India and China.[20] By leveraging private sector partnerships, these projects hope to yield over $120 million in clean-energy financing. A technical assistance program to improve the energy efficiency of televisions in China to achieve Energy Star labeling is expected to result in a decrease of annual CO_2 emissions by 17.7 million tons—the equivalent of removing three million automobiles from the road.[22] In addition, the APP has organized a number of working groups, such as the Cleaner Fossil Energy Task Force, to improve energy efficiency and environmental performance in the use of fossil fuels. This task force is intent on sharing best practices and eliminating perceived market barriers. Its activities are focused on CO_2 storage, post-combustion capture, oxy-firing technologies, coal gasification, and natural gas markets and handling improvements.

The APP is one of the best examples of attempts to create international public-private partnerships that focus on 'clean development' initiatives. It is a voluntary and cooperative effort and provides a new model for initiating carbon reduction projects. Regardless, it has a broad charter, a mixed set of initiatives, and is being co-funded with central government grants whose funding can be selective, intermittent, and even withdrawn. While the APP can be viewed as a facilitator and conduit for new projects, it will provide only a few tens millions of dollars in an effort to resolve a multi-trillion dollar problem. As a result, the impacts of APP initiatives in the near term are likely to be both localized and limited.

REGIONAL INITIATIVES IN THE U.S.

In regard to greenhouse gas emissions and carbon reduction, the motivation of U.S. states is to form alliances to develop a policy platform that has yet to be addressed by the federal government. Their regional initiatives are to develop defined policies and programs in the U.S. that engage groups of states to work in concert to address larger issues of common interest. These policies when implemented can be more effective than states or localities addressing common issues individually—greater resources can be made available when larger populations and their geographic areas are involved. For example, states can pool resources to promote research and fund development projects. There are also benefits to standardizing policies and programs across state borders.

Regional Greenhouse Gas Initiative (RGGI)

On December 20, 2005, the governors of seven Mid-Atlantic and Northeastern states (Connecticut, Delaware, Maine, New Hampshire, New Jersey, New York, and Vermont) announced the creation of the Regional Greenhouse Gas Initiative (RGGI). The governors signed a Memorandum of Understanding agreeing to implement the first mandatory cap-and-trade program for CO_2 in the U.S. The RGGI is establishing a cap on the emissions of carbon dioxide from power plants, and permits the trading of emissions allowances. The program caps emissions at current levels in 2009, and then requires a 10% reduction of emissions by 2019.[23] Pennsylvania and the District of Columbia are observers in the RGGI process.

Massachusetts Governor Deval Patrick committed his state to join the RGGI. In his State of the State address on January 30, 2007 Governor Donald Carcieri announced that Rhode Island would also be joining the RGGI.[24] And in April 2006, Maryland Governor Robert L. Ehrlich Jr. signed into law the Healthy Air Act that required the Governor to include the state in the RGGI by June 30, 2007.[25] Maryland became the 10th participating state in April 2007 with Governor Martin O'Malley's signing of the RGGI Memorandum of Understanding.[26]

The participating states issued a document called the Model Rule on August 15, 2005 (updated on January 5, 2007) detailing the specifics of the RGGI and forming the basis of the individual state regulatory and statutory proposals needed to implement the program.[27] An action plan, updated in August 2006, outlines actions to be taken for program implementation and defines requirements for program support.

To make this a reality, most states were pressed to pass enabling legislation to allow there their states to engage in the RGGI. This represented a difficult set of tasks as the patchwork of existing state statutes needed to be amended to facilitate compliance. Examples of the legislation enacted or being enacted by the individual states includes:

- Connecticut: Public Act No. 07-242, "An Act Concerning Electricity and Energy Efficiency"

- Delaware: SB 263, "An Act to Amend Title 7 of the Delaware Code Relating to a Regional Greenhouse Gas Initiative and CO_2 Emission Trading Program"

- Maine: LD 1851, "An Act to Establish the Regional Greenhouse Gas Initiative Act of 2007"

- Maryland: Subtitle 26.09, Maryland CO_2 Budget Trading Program (draft as of October 2007)

- Massachusetts: Proposed Revisions to 310 CMR 7.29; SB 2768, "The Green Communities Act of 2008"

- New Hampshire: HB 1434, "An Act Relative to the Regional Greenhouse Gas Initiative and Authorizing Cap-and-Trade Programs for Controlling Carbon Dioxide Emissions"

- New York: Draft Part 242, "CO_2 Budget Trading Program"

- Rhode Island: H5577, "An Act Relating to Health and Safety—Implementation of the Regional Greenhouse Gas Initiative Act"

- Vermont: Title 30, Chapter 5, §255, "Regional Coordination to Reduce Greenhouse Gases"

It may be argued that a 10% reduction of CO_2 emissions by power plants in the Northeast is a marginal goal. Since the emissions cap is set at 188 million tons of CO_2 annually which is higher than the current level, the plan can be viewed as having little near-term impact. Regardless, the development of the RGGI to facilitate this goal is truly remarkable. A group of individual states, motivated by a belief that climate change must be addressed, willingly chose to adopt a set of requirements that restricts their rights to generate an important greenhouse gas. They also accepted the likelihood of economic trade-offs. The result is that the RGGI became first mandatory, free-market, cap-and-trade regime for carbon emissions in the U.S.[28]

The RGGI's first auction of carbon allowances occurred in September 2008, raising $40 million for the six participating states (Connecticut, Maine, Maryland, Massachusetts, Rhode Island and Vermont).[29] Many of the qualified bidders represented generators of electrical power. Each allowance sold for an average of $3.07 per ton of excess carbon emissions.[30] The funds netted from the auction are being used primarily for energy efficiency programs, energy conservation and renewable energy projects. Maryland plans to use a portion of its funds to aid low-income families with utility bills. The second auction, held in December 2008, raised $106.5 million at a clearing price of $3.38 per allowance with ten states participating. Massachusetts gained $14.8 million from the sale which will be used for utility energy efficiency programs, heating system upgrades for the homes of low income families, and a "green communities" initiative.[31] Funds from the auction will also support a $5 million training program for energy auditors, insulation installers, and other energy efficiency technicians, creating new "green collar" employment opportunities.[32]

Western Climate Initiative (WCI)

In February 2007, Governors Napolitano of Arizona, Schwarzenegger of California, Richardson of New Mexico, Kulongoski of Oregon, and Gregoire of Washington signed the initial agreement that established the Western Climate Initiative (WCI).[33] This joint effort to reduce greenhouse gas emissions and address climate change was soon adopted by Utah, Montana and the Canadian provinces of British Columbia, Manitoba and Quebec. WCI members agreed to jointly establish regional emissions targets and establish, by August 2008, a market-based, cap-and-trade program covering multiple economic sectors. Many of the policy requirements of the WCI evolved from two previously existing regional agreements: 1) the 2006 Southwest Climate Change Initiative, which included Arizona and New Mexico; and 2) the 2003 West Coast Governors' Global Warming Initiative which included the states of California, Oregon, and

Washington. Idaho, Wyoming, Colorado, Nevada, and Alaska were the initial observers to the WCI process.[34]

In order to implement the WCI, each member state needed to obtain a political consensus and sort out approaches to policy. According to Utah's governor, Jon Huntsman, Jr., "we created the blue-ribbon advisory council on climate change... and came up with 70 different policy approaches."[35] Utah established a policy goal to improve the state's energy efficiency by 20% by 2015. Programs that have already been implemented include converting 7,000 state vehicles to operate using natural gas, establishing a four-day work week for state employees and beginning the process of base-lining and measuring the state's carbon footprint.

In August 2007, the WCI announced a regional, greenhouse gas emissions target of 15 percent below 2005 levels by 2020.[36] The regional target is designed to be consistent with existing targets set by individual member states.

Emissions considered in the WCI included the six primary greenhouse gases identified by the United Nations Framework Convention on Climate Change: carbon dioxide, methane, nitrous oxide, hydrofluorocarbons, perfluorocarbons, and sulfur hexafluoride. Emission reduction actions are intended to be comprehensive and include emissions resulting from stationary sources, energy supplies, waste management, and forestry plus the residential, commercial, industrial, transportation, and agricultural economic sectors.[37] The WCI also requires partners to the agreement to develop action plans, participate in a climate registry, and report on their actions and greenhouse gas inventories every two

Table 4-1. State and Provincial Goals for GHG Reductions

	Short Term (2010-12)	Medium Term (2020)	Long Term (2040-50)
Arizona	not established	2000 levels by 2020	50% below 2000 by 2040
British Columbia	not established	33% below 2007 by 2020	not established
California	2000 levels by 2010	1990 levels by 2020	80% below 1990 by 2050
Manitoba	6% below 1990	6% below 1990[2]	not established
New Mexico	2000 levels by 2012	10% below 2000 by 2020	75% below 2000 by 2050
Oregon	arrest emissions growth	10% below 1990 by 2020	>75% below 1990 by 2050
Utah	Will set goals by June 2008		
Washington	not established	1990 levels by 2020	50% below 1990 by 2050

Source: Western Climate Initiative, Statement of Regional Goal (2007, August 22). http://www.westernclimateinitiative.org/ewebeditpro/items/O104F13006.pdf, accessed 24 August 2008. p. 4.

years. Table 4-2 compares WCI individual and compiled partner goals for the year 2020 and provides the percentage reductions below historical emissions or projected business-as-usual (BAU) levels.

Table 4-2. Summary Compilation and Comparison of 2020 Goals

	Goals					1990-2020 BAU growth
	Relative to 1990	Relative to 2000	Relative to 2005	Relative to 2020 BAU[b]	Absolute Reductions from BAU (MMtCO2e)	
Arizona	35%	0%	-11%	-45%	72	144%
British Columbia	-9%	-27%	-30%	-46%	40	69%
California	0%	-10%	-14%	-28%	170	40%
Manitoba	-6%	-16%	-17%	TBD	TBD	TBD
New Mexico	14%	-10%	-14%	-31%	28	65%
Oregon	-10%	-29%	-32%	-44%	40	61%
Washington	0%	-16%	-11%	-28%	33	40%
Total	2%	-12%	-16%[c]	-33%[d]	383[d]	54%[d]

Source: Western Climate Initiative, Statement of Regional Goal (2007, August 22). http://www.westernclimateinitiative.org/ewebeditpro/items/O104F13006.pdf, accessed 24 August 2008. p. 6. Estimates are as of July 2007.

The methodology used for BAU projections to estimate changes in future CO_2 emission levels is an accepted practice. This approach uses trend analysis which compares average changes in trends over a selected period of years, in this case 1990 through 2005. Trends are charted and projected using historical data to baseline future reductions. Inconsistencies can occur when data is incorrect or unavailable. The greater the number of years included in the projection, the greater the probability that the future projection will be inaccurate and that actual results will vary. The data used and the projected BAU emissions based on trend analysis provide only one case for projecting changes in carbon emissions. Figure 4-1 (using 2005 as the reference year) shows the historical and projected BAU greenhouse gas emissions and compares the projected trend to the changes required to meet a 15% reduction by 2020.

Energy Security and Climate Stewardship Platform for the Midwest

This regional initiative was endorsed by the governors of Illinois, Indiana, Iowa, Kansas, Michigan, Minnesota, Ohio, South Dakota, Wisconsin, and the Canadian province of Manitoba in November 2007 with a goal to "maximize the energy resources, economic advantages

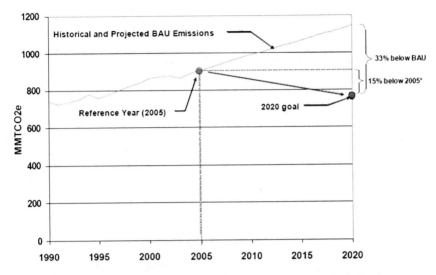

Figure 4-1. WCI Partner GHG Emissions and Regional Goals

Source: Western Climate Initiative, Statement of Regional Goal (2007, August 22). http://www.westernclimateinitiative.org/ewebeditpro/items/O104F13006.pdf, accessed 24 August 2008. p. 5.

and opportunities of Midwestern states while reducing emissions of atmospheric CO_2 and other greenhouse gases."[38] Six of the states (Illinois, Iowa, Kansas, Michigan, Minnesota, and Wisconsin) and Manitoba agreed to establish regional greenhouse gas emission reduction targets with a long-term target of reducing emissions by 68-80%. To reach this goal they agreed to the following objectives:[39]

- Achieve continuous improvement in levels of cost-effective energy efficiency across their economies.

- Comprehensively deploy lower-carbon renewable and fossil fuels and technologies.

- Implement geologic CO_2 storage, terrestrial carbon sequestration and technological utilization of CO_2 on a large scale.

- Add economic value and high-paying jobs to the Midwest's energy, agriculture, manufacturing, and technology sectors by developing and deploying lower-carbon energy production methodologies and technologies.

To implement these objectives, the stated policy options to be explored include quantifying goals for energy efficiency, undertaking state energy efficiency assessments, maximizing the potential of wind energy, removing financial disincentives for energy efficiency improvements, and strengthening building codes.[40] Member states adhering to the platform agreed to have a regional framework in place by 2010. Other states that have adopted portions of the platform include Nebraska and North Dakota. The resolution also establishes a Carbon Management Infrastructure Partnership, a Midwestern Bio-based Product Procurement System, a cross-regional program for biofuels development, and a working group to pursue a multi-jurisdictional transmission initiative.[41]

STATE PROGRAMS IN THE U.S.

Despite the past reluctance of the U.S. federal government to take broad-based initiatives on climate change, local and state governments are moving ahead. There are 16 states, plus the District of Columbia, Puerto Rico, and Guam that require greenhouse gases to be regulated like other atmospheric pollution.[42] Developers of large projects in Massachusetts are now required to "undergo state environmental reviews to assess how they contribute to global warming."[43] State budgets for energy efficiency projects vary widely among U.S. states. Energy efficiency investments in 2006 were $22.54 per resident in Vermont, $.01 per resident in Maryland and averaged $8.50 per resident nationwide.[44]

Policy changes within states tend to occur incrementally. Some states have established renewable portfolio standards (RPS) and energy efficiency portfolio standards, regulations that mainly apply to electrical energy producers in regulated markets. Renewable portfolio standards establish minimum levels of electricity that must be produced by regulated utilities from renewable energy sources such as wind, biomass, solar, hydropower and others. By August 2008, 29 states had mandatory RPS regulations and 5 additional states had functional RPS via voluntary utility commitments.[45] According to Utah's Governor, Jon Huntsman, Jr., Utah "enacted an unprecedented renewable energy portfolio standard" that requires the state's utilities to generate 20% of their total energy requirements using renewable fuels by 2025.[46] New Mexico established its RPS in 2004 and the reductions required are base-lined against 2005 production. It requires rural electric cooperatives to generate 10% of their power from renewable sources by 2020. The larger public electrical producers must produce 20% of their power from renewable sources by 2020 and in addition, must achieve a 5% reduction in electrical energy demand by implementing conservation initiatives.[47]

In California, legislative actions such as the Global Warming Solutions Act of 2006 (AB32) and the emissions performance standards for retail providers of electricity (SB 1368) required unprecedented investments in energy efficiency. Recent updates to the state's Energy Action Plan by the California Public Utilities Commission and the California Energy Commission established goals for meeting energy needs and emphasized energy efficiency, conservation and demand response initiatives. The updated plan called on utility companies to offer comprehensive packages and strategies to save energy. In addition to other approaches, "the update recommends understandable and transparent dynamic-pricing tariffs and demand response programs, time-differentiated default rates for large users, the pairing of advanced meters with automatic infrastructure and dynamic-pricing tariffs from public utilities."[48] Additionally, the California Solar Initiative was passed in 2006—becoming the nation's most comprehensive solar energy policy. It allocates $3.2 billion for energy rebates over an 11 year period.[49]

The California Global Warming Solutions Act of 2006 established the goal of reducing GHG emissions to 1990 levels by 2020. This reduction equates to a 25% reduction from present levels and will likely create a multi-billion dollar carbon market in the state. It further requires reductions of 80% below 1990 levels by 2050.[50] The regulation requires California's Air Resources Board (ARB) to be responsible for monitoring and reducing GHG emissions and to adopt a plan by January 2009 to "achieve the maximum technologically feasible and cost-effective reductions in GHGs, including provisions for using both market mechanisms and alternative compliance mechanisms."[51] Before imposing mandates, the ARB must evaluate economic and health impacts, electrical reliability, conformance with environmental regulations and the effects on low income communities.[52] Other states that have adopted California's limits on GHGs include Connecticut, Maine, Maryland, Massachusetts, New Jersey, New York, Oregon, Pennsylvania, Rhode Island, Vermont and Washington.[53]

Successfully meeting emission reduction targets will require relevant policies and programs. State governments in the U.S. are taking action toward this end. Connecticut has passed a law requiring auto-

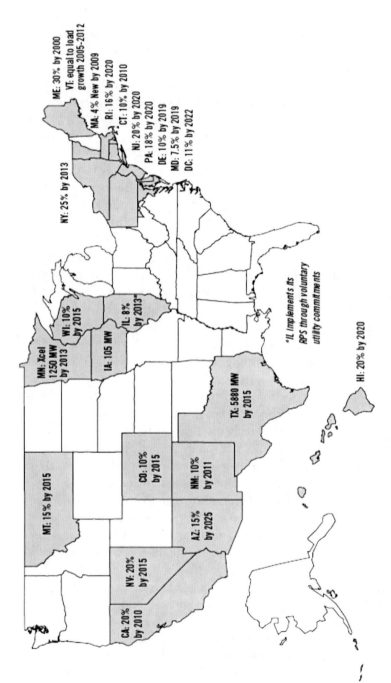

Figure 4-2. State Renewable Portfolio Standards

Source: Pew Center on Global Climate Change. http://www.pewclimate.org/docUp-loads/101_States.pdf, accessed 22 October 2007.

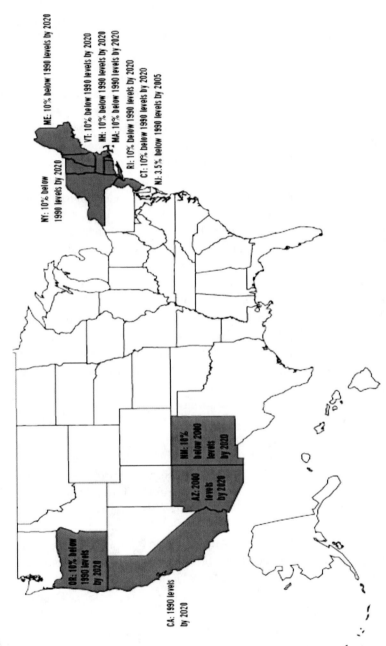

Figure 4-3. State Greenhouse Gas Emission Targets

Source: Pew Center on Global Climate Change. http://www.pewclimate.org/docUploads/101_States.pdf, accessed 22 October 2007.

mobiles sold in 2008 and beyond, to emit less greenhouse gases. Hybrid vehicles that achieve a minimum of 17 km/l (40 mpg) are exempted from Connecticut's 6% sales tax.[54] There are 20 communities in the state that have committed to obtain at least 20% of their electrical power from renewable sources by 2020; these communities have earned the designation as being a "Clean Energy Community."[55] Despite these efforts, Connecticut continues to face other significant challenges such as waste reduction. Currently only 30% of its wastes are being recycled.

The state of New York, through its public service commission, has established an aggressive policy to reduce electric usage by 15% by 2015. This policy hopes to reverse a trend that projects electric usage to increase by 11% for the period. According to Commission Chairman Gary Brown, "Never before have we faced such significant energy challenges... the recent unprecedented rise in energy prices puts to rest any doubts that the market is changing."[56] This policy is anticipated to reduce electrical energy costs by $4 billion and provide additional benefits including employment opportunities. The policy works in combination with programs such as the state's energy efficiency portfolio standard (EEPS) that establishes energy efficiency targets for regulated utilities, requiring them to collect additional set-asides of $172 million annually. The program provides incentives for retrofitting outdated and inefficient residential, commercial and industrial properties with energy efficient equipment. Other programs active in New York include a renewable energy portfolio standard, a regional greenhouse gas initiative, and improvements in building codes along with appliance energy standards.[57]

States are finding creative ways to reduce their fossil fuel-based energy usage. Hawaii is 90% dependent on carbon-based fuels. About 80% of Hawaii's energy is from diesel fuel that is used to generate electricity. The state has a goal of providing 70% of its energy from renewables by 2030 and has adopted the Hawaii Clean Energy Initiative in an effort to meet this goal. Electric water heaters account for 25% or more of the electrical energy usage in residences. Hawaiian Electric estimates that 85,000 homes, roughly 20%, already have solar hot water heating systems. In 2008, Hawaii passed legislation that requires all new homes to have such systems beginning in 2010.[58] The state is also developing wind farms, yet in some areas electrical infrastructure is lacking. Despite the availability of wind power, wind generators on the island of Maui are often disabled due to fears of overloading the island's electrical grid.

Kansas is another state that is choosing to develop its extensive wind resources. Coal plants require 7-10 years to approve, design and construct. After rejecting a permit to build a coal-fired power plant in 2007 citing unmanageable carbon dioxide emissions, Kansas offers a heuristic example of how quickly wind resources can be developed. Wind farms that have been installed in the past include: 1) Westar Wind, 1.5 MW developed in 1999; 2) the Clay County Wind Energy facilities in southwestern Kansas, 112 MW developed in 2001; 3) the Elk Ridge Wind Farm, 150 MW developed in 2005; and 4) the Spearville Wind Energy Facility, 100 MW developed in 2006. As opposed to meeting electrical energy requirements with fossil fuel-fired power plants, the state embarked on a cooperative initiative with its public utilities to expand wind power capacity. During September 2008 Lieutenant Governor Mark Parkinson proudly announced that Kansas had met its goal of having over 1,000 MW in wind generating capacity, making the state the third-largest in terms of capacity and a state that tripled wind capacity in only three years.[59] These wind farms are being developed without financial incentives from the state government and comprise a total of 650 MW of additional electrical generation capacity.[60] The newly developed sites include:

Smoky Hills Wind Farm—a two-phase development of 250 MW in Lincoln and Ellsworth Counties which is the largest commercial wind farm in Kansas.
Meridian Way Wind Farm—201 MW in Cloud County.
Central Plain Wind Farm—99 MW in Wichita County.
Flat Ridge Wind Farm—100 MW in Barber County.

The Meridian Way Wind Farm features the new 3 MW Vestas V-90 turbines, among the largest machines to be installed on land-based wind farms. These mammoth machines have a diameter of 90 meters (295 feet) and reach rated output with wind speeds of 15 m/s (50 ft/s). Oregon offers yet another example. The state now has over 1,000 MW of installed wind generation capacity and an additional 1,000 MW under development.

States are also changing their service schedules. Utah, in 2008, became the first state to institute a mandatory four-day workweek. The change applies to 17,000 employees and is being instituted to reduce the state's carbon footprint, increase energy efficiency, save money and provide workers fewer commutes and more flexibility.[63] As a symbol

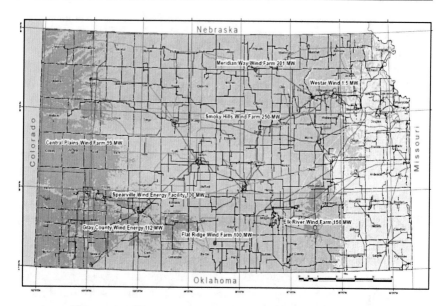

Figure 4-4. Site Locations of Wind Farms in Kansas[61]

Figure 4-5. Vestas V-90, 3 MW Wind Turbines (Meridian Way Wind Farm, Kansas)[62]

of the state's commitment, the governor's mansion, which Governor Huntsman calls "the best public housing in the world," is now powered by wind power.[64] According to the governor, "the reason that we're doing this is obviously energy efficiency and reducing our overall carbon footprint."[65]

SUMMARY

Since global climate change surfaced on the public policy agenda in the late 1900s, governments have passed environment-friendly legislation. Such legislation identifies increases in GHG and CO_2 emissions as a primary cause for global warming. This chapter stresses the importance of government legislation, policies, and programs in the effort to avoid global climate change. Such legislative frameworks include economic, political and social incentives that encourage or mandate change. In the past, much legislation focused on increasing energy resources and encouraging economic development. However, as the impact of high atmospheric carbon concentrations became evident and a scientific consensus developed, governments have redirected their policies. Now, many government initiatives focus on carbon reduction strategies and technologies.

This chapter acknowledges that without hierarchical coordination of the levels of governance (e.g., state, national, and international), mitigation policies are bound to vary not only between nation states, but also within them. Global climate change is a universal, nondiscriminatory threat that has the potential to severely change the Earth's ecosystems. Indeed, the effects of climate change will be widespread; they will not distinguish between the governments that made huge efforts to cut carbon emissions, those who barely contributed and those most responsible. Despite the universal threat, governments show surprising inconsistencies in their efforts to reduce carbon emissions and avoid global climate change. International treaties, such as the Kyoto Protocol, permit such discrepancies through flexible mechanisms and selective participation. The array of government policies, strategies and technologies demonstrates the overall complexity of mitigation efforts.

Increasingly, state governments in the U.S. are being recognized as leaders in carbon reduction strategies. Some states have passed more stringent legislation than the U.S. government, surpassing the standards set in international treaties. State governments or city councils are more readily accountable to citizens, lobbyists, and special interests groups concerned about global warming. Consequently, many states have established groups, teams and even departments to monitor and implement carbon reduction strategies. One advantage of state initiatives is that they are specifically tailored to a particular geography, population, climate, and economy. State measures contextually match their areas and populations and frequently have a more effective impact than national or international measures.

Governments use different policy mechanisms to advance their carbon reduction agenda. Some governments work in concert with others and some work internally. International public-private partnerships such as the Asia-Pacific Partnership offer commercial and industrial concerns the opportunity of voluntary participation. Regional initiatives in North America include the Regional Greenhouse Gas Initiative (RGGI), the Western Climate Initiative and the Energy Security and Climate Stewardship Platform for the Midwest. Individual states are establishing legislation, such as the California Global Warming Solutions Act of 2005 that mandates reductions in future carbon emissions. Other states are having success with programs to support alternative energy development, energy conservation and energy efficiency.

Endnotes

1. Energy Information Administration. *(2004) India: Environmental issues* and *China environmental issues (2003)*, Country analysis briefings.
2. Vigar, D. (2006). *Climate change: The role of global companies.* London: Tomorrow's Company.
3. John G. (2008, January 13). Nuclear energy renaissance worries some observers. *The Courier-Journal.* p. A9.
4. *Ibid.*
5. *Ibid.*
6. Commonwealth of Australia, Department of Climate Change (2008, July). *Carbon pollution reduction scheme green paper.* p. 8.
7. *Ibid.,* p. 14.
8. *Ibid.,* p. 12.
9. Herd-produced methane poses a challenge (2008, June 30). *The Courier-Journal.* p. E3.
10. *Ibid.*
11. Card, R. (2003, December 11). *Public-Private Partnership for Technology Innovation.* Remarks by the Under Secretary of Energy (U.S) at a roundtable discussion on Technology, Including Technology Use and Development and Transfer of Technologies. http://www.state.gov/g/oes/rls/rm/2003/27102.htmhttp://www.state.gov/g/oes/rls/rm/2003/27102.htm, accessed 13 September 2008.
12. *Ibid.*

13. *Ibid.*
14. Department of State Publications (2007). *Asia-Pacific Partnership on clean development and climate.* p. 17
15. *Ibid.,* p. 22
16. *Ibid.,* p. 2-16
17. *Ibid.*
18. As quoted from his a presentation at the NASEO Annual Meeting, Overland Park, Kansas on 9 September 2008. Ambassador Harnish is the Principal Deputy Assistant Secretary, Bureau of Oceans and International Environmental and Scientific Affairs, U.S. Department of State.
19. *Ibid.*
20. *Ibid.*
21. *Ibid.*
22. *Ibid.*
23. Pew Center (2008, July 10). *Regional initiatives.*
24. *Ibid.*
25. *Ibid.*
26. *Ibid.*
27. Regional Greenhouse Gas Initiative. *Model rule and memorandum of understanding.* http://www.rggi.org/modelrule.htm., accessed 24 August 2008.
28. Energy Vortex (2009). *Northeast greenhouse gas allowance auction raises $106.5 million.* http://www.energyvortex.com/pages/headlinedetails.cfm?id=3966, accessed 10 February 2009.
29. Shulman, R. (2007, September 30). Carbon sale raises $40 million. *The Washington Post.* p. A4.
30. *Ibid.*
31. Energy Vortex (2009). *Northeast greenhouse gas allowance auction raises $106.5 million.* http://www.energyvortex.com/pages/headlinedetails.cfm?id=3966, accessed 10 February 2009.
32. *Ibid.*
33. Pew Center on Global Climate Change (2008, July 10). *Regional initiatives.*
34. *Ibid.*
35. Huntsman, J. (2008, September 19). In a keynote address at the *Energizing Kentucky Conference,* Louisville, KY.
36. Pew Center on Global Climate Change (2008, July 10). *Regional initiatives.*
37. Western Climate Initiative (2007, August 22). *Statement of regional goal.* http://www.westernclimateinitiative.org/ewebeditpro/items/O104F13006.pdf, accessed 24 August 2008.
38. MGA Energy Summit (2007, November 15). *Energy security and climate stewardship platform for the Midwest.* http://64.233.167.104/search?q=cachersDdyQ2Sx0J:www.nationalwind.org/pdf/MGA_Platform.pdf+Climate+Stewardship+Platform+for+the+Midwest&hl=en&ct=clnk&cd=4&gl=us, accessed 24 August 2007. p. 4.
39. *Ibid.*
40. *Ibid.,* p. 7.
41. Pew Center on Global Climate Change (2008, July 10). *Regional initiatives.* http://www.pewclimate.org/what_s_being_done/in_the_states/regional_initiatives.cfm., accessed 24 August 2008.
42. Ritter, J. (2007, June 6). California sees sprawl as warming culprit. *USA Today.* p. 1.
43. *Ibid.*
44. Zinga, S. (2008, September 18-19). Presentation at the Energizing Kentucky Confer-

ence, Louisville, KY.
45. Pew Center offers an interactive map of RPS regulations that is updated periodically and available at http://www.pewclimate.org/what_s_being_done/in_the_states/rps.cfm.
46. Huntsman, J. (2008, September 19). In a keynote address at the *Energizing Kentucky Conference*, Louisville, KY.
47. Martinez, F. (2008, September 8). Presentation at the NASEO Annual Conference. Overland Park, KS.
48. Energy Vortex. *CPUC release updated Energy Action Plan focused on climate change.* http://www.energyvortex.com/pages/headlinedetails.cfm?id=3416, accessed 16 March 2003.
49. California Public Services Commission (2006, January 12). http://www.cencibel.biz/images/CPUC_Solar_Policy_Jan-12-06.pdf
50. *What is AB32?* (2008, February). http://www.cityoffullerton.com/civica/filebank/blobdload.asp?BlobID=4788, accessed 16 November 2008.
51. California Air Resources Board (2006, September 25). *AB32 fact sheet—California Global Warming Solutions Act of 2006.* http://www.arb.ca.gov/cc/factsheets/ab-32factsheet.pdf, accessed 20 October 2007.
52. *Ibid.*
53. Healey, J. (2007, September 13). Judge says states can regulate emissions. *USA Today.* p. B1.
54. Holahan, D. (2007, November). Just warming up. *Connecticut Magazine.* p. 76.
55. *Ibid.*, p. 77.
56. Historic energy efficiency program gets underway in New York. (2008, 19 June). *Kentucky energy watch.* 9 (25).
57. New York State Public Service Commission (2008, July 19). *Historic energy efficiency program gets underway in NY.* Tdworld.com/customer_service/ny-energy-efficiency-0806, accessed 21 July 2008.
58. Hawaii acts to require solar water heaters. (2008, June 30). *The Courier-Journal.* p. D2.
59. As stated by the Lieutenant Governor, Mark Parkinson, at a luncheon in Overland Park, KS on 2 September 2008.
60. Kansas Corporation Commission (2008, April 11). *Kansas wind resources map.* http://www.kcc.state.ks.us/energy/kswindmap.pdf, accessed 13 September 2008.
61. *Ibid.*
62. Vestas. *Number 1 in Modern Energy.* http://www.vestas.com/en/wind-power-solutions/wind-turbines/3.0-mw, accessed 13 September 2008.
63. Copeland, L. (2008, July 1). State workers in Utah shifting to 4-day week. *USA Today.* p. 2A.
64. Huntsman, J. (2008, September 19). In a keynote address at the *Energizing Kentucky Conference*, Louisville, KY.
65. KCPW (2006). *Governor's Mansion, SUV go green.* http://www.kcpw.org/article/3674, accessed 16 October 2008.

Chapter 5

Local Carbon Reduction Policies

"Climate change may be the most critical issue we face today. That may seem like a dramatic statement, but all science points to catastrophic results if we don't act quickly to get a handle on this growing crisis. It's a global problem, but it's a local problem too and our responses have to be both global and local. Austin has long been a national leader on energy efficiency, renewable power and innovative technologies. Now we need to push those efforts to the next level."

Will Winn, Mayor of the City of Austin (2007)[1]

"Climatically, it means nothing... if all the nations of the world lived up to the Kyoto Protocol the effect on global warming would be undetectable for a century. So the effect of a few cities within the United States living up to Kyoto would be less than undetectable."

Patrick Michaels, Cato Institute[2]

An overriding determinate of the formation cities has been the availability of inexpensive energy supplies (Hough 1995:16). Energy is needed to sustain the industrial, residential, commercial and transportation needs of urban environments. Cities are structured in a way that mimics natural systems, ingesting supplies of materials and energy, processing them, providing services and generating wastes. The energy flows through cities are roughly 100 times greater than the energy processed by natural ecosystems. This concentration of energy uses has inherent systemic inefficiencies that provide opportunities to use carbon-based energy supplies more effectively. Urban and county governments, in particular, have the ability to reduce carbon emissions within their jurisdictions. Local solutions must be sought to foster the health and longevity of the systems that fundamentally support well-being of their localities. Managing the energy use of cities is part of the solution. According to Portney, "Sustainable cities frequently attempt to address energy issues" by directly influencing the city's "consumption of energy." This is done by offering consumers public transportation alternatives and creating home energy conservation opportunities (Portney 2003:95). Promoting an urban ecology that engenders construction of environmentally sensitive developments and energy efficient structures is a valid alternative.

Local policies structure the theoretical basis for local initiatives. Cities have a broad range of options that go beyond simply changing energy consumption and energy production patterns and styles of energy demand management programs. Regarding energy production, environmental impacts can be mitigated with greater use of renewable energy, cogeneration, district heating and cooling systems, and switching from coal use to natural gas (Leitmann 1999:278).

Cities have infrastructure that must be managed. Infrastructure influences sustainability and consumes resources. Cities and local governments often control and influence infrastructure decisions that affect energy use, and they can create long term commitments to the forms of energy utilized to provide public services.

Cities have choices. Since urban areas account for the majority of energy usage, it is clear that our cities must engage in meaningful policy approaches to effectively address carbon reduction. Changing the behavior and policy direction of regimes in cities can be a difficult and complex task.

Cities create policies. Policies impact the sustainability of cities and their energy consumption. There are policies and technologies that cities can employ to reduce energy use, and others that can cause inordinate quantities of energy to be consumed.

Cities have broad powers to control local transportation and other infrastructure decisions. More efficient transportation equipment can also be placed into service. Cities can require their fleets to use natural gas or opt for vehicles that use other forms of alternative energy. Cities can expand their existing mass-transit systems.

Cities own equipment. Equipment consumes energy. Cities purchase computers, office equipment, refrigerators, air conditioners, desk lamps, motorized devices. Cities can choose to purchase the lowest cost equipment item or pay a premium, if necessary, for equipment that uses less energy. Cities can choose to optimize the efficiency of equipment in order to minimize its energy use and reliability—or they can defer maintenance and cause premature equipment failure. Cities can purchase energy-saving equipment to reduce energy use—or they can dismiss energy use as a component in the decision making process.

Cities can tax energy to promote energy conservation. They can tax fuels or carbon emissions in hope of improving air quality, or alternatively, provide subsidies to their residents for the costs of energy.

Cities purchase energy. Purchasers can make demands of energy suppliers. Cities can decide what type of fuels they would prefer to use and require efficiency standards of their suppliers. They can sometimes negotiate favorable and cost-effective utility rate schedules and establish demand-side performance-based standards. Cities can choose to require that public utilities and their officers not be permitted to contribute to local political campaigns. Cities can purchase "green energy" produced by using alternative fuels.

Each city has a carrying capacity which can be difficult to assess. Carrying capacity has been defined as "the ability of a city's surrounding environment to generate resources and assimilate wastes" which impacts the urban quality of life (Leitmann 1999:38). Since carbon emissions are a waste product of urban environments and their environmental impacts are becoming apparent, reducing such emissions is important.

Reducing the CO_2 emissions of cities can be difficult. The development patterns of cities also play a role. Growing cities are expanding their ecological footprints and straining local environmental resources. Ecological footprints are defined as the land and water area that is required to support a defined human population and material standard indefinitely using prevailing technologies (Newman 2008:260). The ecological footprints of cities vary. Cities that consume more carbon-based fuels for energy-consuming processes have larger ecological footprints. Cities with stable populations have an opportunity to improve existing infrastructure and reduce their demand on resources. Often, this is not the case. If their population decreases, resource needs may diminish, yet their infrastructure remains in place and the demand for resources may fail to decline in tandem with population losses. Cities that are in decline often have limited budgets to make infrastructure improvements. When budgets are constrained, transit systems are not upgraded, new technologies are not employed and such cities often begin to deteriorate. In each case, reducing carbon emissions is dependant on the development patterns and the capability of cities to improve their existing infrastructure. Long term policies and programs must be identified and supported in order to meet the challenge of reducing carbon emissions. This requires assessing the local impact and modifying the local infrastructure. Many cities are doing just this.

Regardless, state and city governments are adopting new and exciting policies in their efforts to green their localities. In 2007, 14 states adopted green building policies and in 2008, eight states (including Oklahoma, New Jersey, and South Dakota) and 22 localities endorsed new green initiatives.[3] This trend has been attributed to climate change concerns and higher energy costs. According to Lynn Spruill, the chief administrative officer of Starkville, Mississippi, "We are leading by example… we've got to do something to reduce our negative impact on the environment."[4]

THE LOCAL IMPACT OF CARBON EMISSIONS

The impact of carbon emissions is measured in units of million of metric tons (MMT)[5] of carbon dioxide equivalents. Reducing a ton of CO_2 emissions can be a difficult task. Table 5-1 helps bring perspective to the enormity of the task.

Since mitigation actions occur locally, removing a ton of carbon from the atmosphere or preventing its release has larger implications. Such actions reduce an area's carbon footprint.

LOCAL GOVERNMENTAL POLICIES

The U.S. Mayors Climate Protection Agreement was endorsed at the 2005 Annual U.S. Conference of Mayors in Chicago. Mayors of U.S. cities who signed the Mayor's Climate Change Agreement agreed to meet or exceed the Kyoto Protocol's requirements for reducing global warming.[7] Actions being taken include instituting anti-sprawl land-use policies, developing urban forests, establishing public information campaigns, and encouraging state governments to become more involved.[8]

As of July 2007, over 600 mayors have signed the document; their cities represent a population in excess of 50 million. This agreement has three primary components:[9]

Table 5-1. Equivalents of 1 MMT of CO$_2$ emissions[6]

1 MMT of CO$_2$ is equivalent to:

- Replacing a 500 MW cold-fired power plant with two 500 MW combined-cycle gas-fired power plants and operating them for one year.

- 216,000 passenger cars not driven for one year.

- Reducing gasoline use by 114 million gallons, equal to 13,400 tanker trucks.

- Reducing oil use by 2.3 million barrels.

- Every adult in the state of California walking up one floor each workday rather than taking an elevator.

- The amount of electricity used by 128,000 average U.S. households (or 193,000 California households) in a single year.

- The energy saved by replacing 13 million standard incandescent light lamps with compact fluorescent lamps.

- 26,000,000 tree seedlings grown for 10 years.

- 833,000 acres of pine or fir forest storing carbon for one year.

1) Urges state and federal governments to enact policies and programs intended to reduce pollution that causes climate change below 1990 levels by 2012.

2) Urges the U.S. Congress to pass legislation establishing timetables and GHG emission limits, with a market-based system of tradable allowances among emitting industries.

3) Pledges that the cities involved will take action to meet or exceed Kyoto Protocol targets for reducing global warming pollutants.

Actions to which cities have committed include inventorying greenhouse gases, updating building codes, encouraging use of public

The U.S. Mayors Climate Protection Agreement
(As endorsed by the 73rd Annual U.S. Conference of Mayors meeting, Chicago, 2005)

A. We urge the federal government and state governments to enact policies and programs to meet or beat the target of reducing global warming pollution levels to 7 percent below 1990 levels by 2012, including efforts to: reduce the United States' dependence on fossil fuels and accelerate the development of clean, economical energy resources and fuel-efficient technologies such as conservation, methane recovery for energy generation, waste to energy, wind and solar energy, fuel cells, efficient motor vehicles, and biofuels;

B. We urge the U.S. Congress to pass bipartisan greenhouse gas reduction legislation that 1) includes clear timetables and emissions limits and 2) a flexible, market-based system of tradable allowances among emitting industries; and

C. We will strive to meet or exceed Kyoto Protocol targets for reducing global warming pollution by taking actions in our own operations and communities such as:

1. Inventory global warming emissions in City operations and in the community, set reduction targets and create an action plan.
2. Adopt and enforce land-use policies that reduce sprawl, preserve open space, and create compact, walkable urban communities;
3. Promote transportation options such as bicycle trails, commute trip reduction programs, incentives for car pooling and public transit;
4. Increase the use of clean, alternative energy by, for example, investing in "green tags", advocating for the development of renewable energy resources, recovering landfill methane for energy production, and supporting the use of waste to energy technology;
5. Make energy efficiency a priority through building code improvements, retrofitting city facilities with energy efficient lighting and urging employees to conserve energy and save money;
6. Purchase only Energy Star equipment and appliances for City use;
7. Practice and promote sustainable building practices using the U.S. Green Building Council's LEED program or a similar system;
8. Increase the average fuel efficiency of municipal fleet vehicles; reduce the number of vehicles; launch an employee education program including anti-idling messages; convert diesel vehicles to bio-diesel;
9. Evaluate opportunities to increase pump efficiency in water and wastewater systems; recover wastewater treatment methane for energy production;
10. Increase recycling rates in City operations and in the community;
11. Maintain healthy urban forests; promote tree planting to increase shading and to absorb CO2; and
12. Help educate the public, schools, other jurisdictions, professional associations, business and industry about reducing global warming pollution.

transit systems, supporting alternative energy, promoting LEED construction practices, augmenting recycling programs and promoting tree planting programs to increase carbon dioxide absorption. Land use policies, such as those that promote suburban developments rather than mixed use developments with higher population densities, also create longer commuting distances, increasing the use of fossil fuels.

Suburban Development Near Chicago, Illinois

Portland, Oregon—the first U.S. city with a global warming action plan—is among the cities that have signed the agreement. Portland and Multnomah County established a goal to reduce greenhouse gases to 10% below 1990 levels by 2010.[10] Portland's accomplishments include a 75% increase in the use of public transportation systems (since 1990) and improvements to encourage bicycle use. Without Multnomah County's carbon reduction program, it would be emitting more than 12 million tons of CO_2 annually.[11] Due to Multnomah County's progressive actions, CO_2 emissions have been reduced to an annual total of 9.7 million metric tons.[12] Washington, D.C., replaced 414 diesel buses with ones that use compressed natural gas.[13] Other cities, including Evansville, IN; Lexington, KY; and Baltimore, MD, have installed LED traffic signals that have reduced electrical use by over 80%. Due to their longer lamp life, using LED lamps has lowered the fuel required for signal service and repair. Seattle's municipal government has reduced carbon emissions by 683,000 metric tons annually (60% below 1990 levels) by using hybrid-electric vehicles and reducing fleet fuels—an amount equal to the emissions from 148,000 automobiles.[14]

The U.S. Mayors Climate Protection Agreement has been adopted by cities such as Miami, Los Angeles, Chicago, New York, and many others. Perhaps more impressive are the many smaller cities across the U.S. that have established local initiatives to lessen their impact on climate change. Why are their activities so impressive? Smaller communities often have fewer local resources for specialized initiatives. For them, programs such as these require developing internal resources and skills, and often investing in outside resources as well.

Consider the city of Cambridge, Massachusetts. This city, actually part of the Boston Metropolitan Statistical Area, is sandwiched between the municipalities of Boston, Somerville and Brookline. The metropoli-

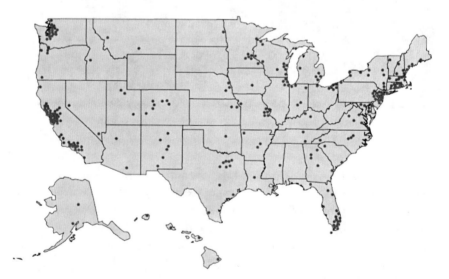

Figure 5-1. Cities Committed to the U.S. Mayors Climate Protection Agreement

Source: Pew Center on Global Climate Change. http://www.pewclimate.org/docUploads/Climate-101-LocalBlueline.pdf, accessed 22 October 2007.

tan area of Boston is home to Boston College, Massachusetts Institute of Technology, Harvard University and other educational institutions. Cambridge has a mass transit system that is part of the Massachusetts Bay Transportation Authority (MBTA), that includes buses, commuter rail, a subway system known locally as the "T" (the oldest in the U.S.) and ferries.[15] This transportation system serves eastern Massachusetts, Logan Airport, and Providence, Rhode Island. It is possible for residents to live in Cambridge without owning an automobile. There are actually disincentives to automobile ownership, such as the scarcity and high cost of parking spaces. On-street parking is available, but limited. Where possible, parking garages are strategically located near subway stations—such as one located just a half block from Harvard Station. This allows commuters to park vehicles near a transit station and provides access to the rail system.

In a city that is almost entirely "built-out," meaning that most of the land has been developed, mixed-use developments are commonplace in Cambridge. Many buildings of historical significance, including many churches, the buildings on the Harvard University campus and the venerable City Hall Building with its signature bell and clock tower, have

Harvard Square in Cambridge, Massachusetts

been restored to their former glory. Development in the city includes medium and high-rise buildings. There are also apartments located above commercial businesses that are located at the street and lower levels. Mixed-use developments allow residents to walk to work and nearby amenities. Such developments provide the population density necessary to allow public transit systems to be successful.

CITIES GOING GREEN

Local governments are experimenting with financial incentives to help "green" their cities. In Parkland, Florida, the city motto is "Environmentally Proud." Residents are being issued checks for installing low-flow toilets or shower heads, replacing air conditioners with more efficient models or purchasing a hybrid automobile. The city predicts that this program will cost $100,000 in its first year. Could this initiative bankrupt the city government? Vice-Mayor Jared Moskowitz comments, "I can only wish that so many residents want to go green that this becomes an issue."[16]

Cities, including Phoenix, Philadelphia, Hartford and Ann Arbor, are creating energy manager positions within their organizations to better manage utility consumption and develop creative programs to reduce energy use.[17] Arlington County Government, that contains Arlington, VA, created a revenue steam with a tax on residential energy use that generates $1.5 million annually to maintain its Fresh AIRE (Arlington Initiative to Reduce Emissions) campaign.[18] The program's goal is to reduce the county's greenhouse gas emissions by 10% by 2012 from year 2000 levels.[19]

Beginning in 2008, San Francisco will offer homeowners maximum rebates of $5,000 for solar panel installations if they choose to use a local contractor. The city will also provide up to 90% of the costs of making apartment buildings more energy efficient.[20] The city has created the San Francisco Carbon Fund—a program for carbon offsets to fund local green initiatives. This program provides the opportunity to align local carbon reduction activities with local sustainability goals. According to Mayor Newson, "Our carbon offset program will achieve meaningful, measurable reductions in greenhouse gas emissions… it is the first effort of its kind, where you can buy carbon-offsets for projects that take place in San Francisco, that directly benefit San Francisco."[21]

The city's initial efforts include an information campaign to advertise the program, providing information on the costs of carbon intensive activities, determining ways to assure the quality of the carbon offsets and issuing a request for proposals (RFP) for local greenhouse gas reduction projects.[22]

Cities are also establishing green building programs. Cities have a wide range of options, and programs vary in their green building initiatives. While some provide general specifications for site and construction practices, others adopt established requirements for their buildings such as LEED and Energy Star. Salt Lake City has required new municipal buildings to achieve LEED Silver certification since 2005. Cities may choose to provide optional specifications for structures in their jurisdictions or establish mandates that are defined in ordinances and codes. Boston has developed its own green building standards and incorporated them into their municipal building code.

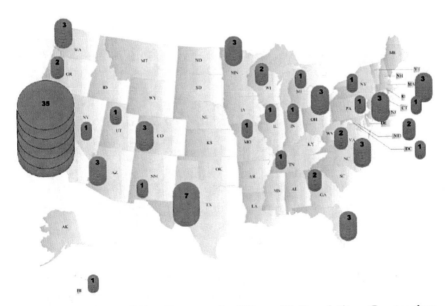

Figure 5-2. Green Building Programs in Cities with Populations Greater than 50,000 (Number of cities in each state are indicated)

Source: Rainwater, B. (2007). *Local Leaders in Sustainability—A study of green building programs in our nation's communities.* American Institute of Architects. http://www.aia.org/SiteObjects/files/LLinSustain(Findings)_Final.pdf, accessed 17 February 2008.

AUSTIN CITY LIMITS ON CARBON

The City of Austin, Texas (with a metro area population of 1.5 million), has a number of exemplary programs and policies that are broadly directed toward reducing energy use and carbon emissions. Austin's programs have been successful and the city continues to be a leader in energy efficiency.

The impact of Austin's city-wide energy conservation program allowed its utility to avoid construction of a planned 500 megawatt power plant that would have been needed as early as 2000. The city is uniquely positioned to implement its programs, as it owns Austin Energy—the electricity utility and the metro area's water supply and water treatment facilities. Austin's policies include: 1) community-wide goals and directives; 2) policies that impact local government facilities and equipment; 3) programs for private businesses, residences and construction practices; and 4) utility programs. This is an example of a policy that is documented as a set of plans—each with a defined set of goals—that are both challenging and comprehensive. These policies and goals are codified in the Austin Climate Protection Plan, announced by Mayor Will Wynn in February 2007, and are structured as five discrete plans:[23]

1) **The Municipal Plan**—creates a goal of making all city facilities, fleets and operations carbon-neutral by 2020. This plan requires the development of policies, procedures, targets and reporting standards to achieve the "reduction of GHG emissions and energy consumption in all city departments."

2) **The Utility Plan**—implements an aggressive GHG reduction program that focuses on renewable energy, energy efficiency improvements, requires carbon-neutral generation and the retirement of utility infrastructure generating GHG emissions. It establishes the goal of achieving 700 MW in savings from the implementation of energy conservation programs and an increase in electrical generation from renewables. More specifically, this plan intends to expand renewable resources, in terms of electricity production, from 6% to 30% by 2020. It also requires that a cap on CO_2 emissions be established.

3) **The Homes and Buildings Plan**—dictates changes in residential and commercial building codes that mandate energy efficiency in new and existing homes and buildings. It requires that new single-family residences be "zero net-energy capable" by 2015, that a carbon neutral certification program be developed, and offers enhanced incentives for green buildings.

4) **The Community Plan**—provides for the development of a comprehensive approach to the reduction of GHGs on a community-wide scale. This includes the establishment of a City Climate Action Team to inventory GHG emissions, develop long-term reduction requirements and implement strategies for the metropolitan area.

5) **The Go Neutral Plan**—offers mechanisms for businesses and individuals to reduce their carbon footprint. This program will promote carbon neutrality by creating: a set of local GHG reduction strategies for citizens, businesses and organizations; an online carbon footprint calculator; and mechanisms for the purchase and exchange of carbon offset credits.

The City of Austin plans to make its facilities more energy efficient and power them with renewable energy (primarily wind, solar and biomass) by 2012 and has a goal of becoming carbon neutral by 2030. To this end, the City's 10,000 employees will be given global warming and outreach education and training.[24] Each city department is tasked with developing and implementing a climate protection plan. The City of Austin Water Utility, the department that manages water resources and wastewater, is challenged to reduce electricity and water usage in the city. Austin Water Utility hopes to avoid constructing new treatment facilities. This is particularly important since the department accounts for half of the electricity consumed by government facilities.

The city also has a goal of making all government vehicles carbon neutral by 2020. Austin currently has 59 hybrid vehicles and six garbage trucks that use compressed natural gas.[25] The city intends to have the entire fleet of city-owned vehicles, including heavy equipment, powered by electricity or non-petroleum fuels to the extent that it is technically possible.[26]

COMMUNITY CARBON REDUCTION PROJECT

of Cambridge, England, developed an ambitious goal in 2003 of reducing its carbon emissions by 60% by 2025. To achieve this goal, the city collaborated with the Community Carbon Reduction (CRed) program housed at the University of East Anglia in Norwich and the University of North Carolina in Chapel Hill. The action plan developed involves inventorying carbon emissions, directing resources to effectively lower their emissions plus a citywide approach to emissions monitoring. The approach involves a six-step plan:[27]

1) The sectors of energy use in the city are identified: residential, business, industry, transport, municipal operations, university and colleges.

2) For each sector, a carbon dioxide inventory is created showing the contribution of that sector to the overall city emissions. The contribution of each form of energy use in that sector is evaluated to determine the total emissions from that sector (e.g., the fraction of emissions from residential due to space heating).

3) Strategies are next developed to reduce energy use and the emissions from each sector, focusing attention on strategies that would yield the greatest reduction in carbon dioxide emissions (e.g., loft insulation in existing homes).

4) Individuals and institutions are identified for each sector to serve as the points of contact between CRed and that sector. CRed will help move reductions forward in that sector (e.g., developers in the case of the residential sector).

5) Resources are created to help these individuals or institutions implement the strategies (e.g., information on tax credits, or providing free energy audits).

6) The carbon dioxide emissions inventory is updated periodically to measure progress towards the emissions reduction goal.

Specific strategic goals that have been identified include reducing energy use in municipal operations by 25% and reducing residential emissions by 5%. To achieve this the city is promoting waste minimization, reducing emissions by municipal transport and staff travel, creating awareness programs and advancing sustainable design practices in construction.

RANKING CITIES BASED ON THEIR CARBON EMISSIONS

Carbon dioxide emissions vary widely across major U.S. metropolitan areas. Interestingly, the per capita carbon footprint of those living in U.S. cities is 14% less than those who live in non-urban areas.[28] The use of carbon-based transportation fuels and residential energy use have been proven to be statistically correlated to urban sustainability.[29] This is likely due to higher population densities and greater transportation options. Recently, a report called *Shrinking the Carbon Footprint of Metropolitan America* from the Brookings Institution quantified the carbon footprint of the country's 100 largest metropolitan areas based on mobility fuels and energy use in residential structures—factors that contribute to about half of total carbon emissions.[30] The report indicated that metro areas with "high density, compact development and rail transit offer more energy and carbon efficient emission lifestyles" than urban areas that tended to have more auto-centric, sprawling development patterns.[31] The report also indicated that while the population of metro areas increased 6.3% during the most recent study period, the average per capita carbon footprint of the 100 metro areas grew by only 1.1%, significantly less than the growth of carbon emissions in other non-metro areas.[32] There is broad variance in the carbon footprints of individual U.S. cities.

Regional differences in per capita carbon footprints are apparent—west coast cities typically generate far less carbon than those located in the Midwest, Southeast and Atlantic Seaboard. In the northeast, many cities have higher than average per capita carbon emissions due to their reliance on fuel oil for residential heating. Per capita annual carbon emissions from transportation and residential energy use is lower in cities such as in Honolulu (1,156 metric tons), Los Angeles (1,416 metric tons) and New York City (1,495 metric tons), but emissions are much higher in cities such as Lexington, KY (3,455 metric tons), Indianapolis (3,364 metric tons) and Cincinnati (3,281 metric tons).[34] Higher emissions are likely due to differences in climate, public transportation availability

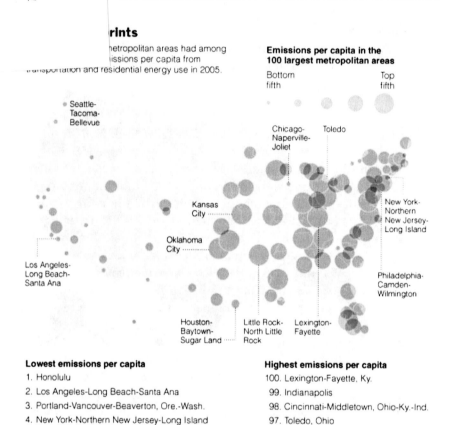

rints

etropolitan areas had among
issions per capita from
transportation and residential energy use in 2005.

**Emissions per capita in the
100 largest metropolitan areas**

Bottom
fifth

Top
fifth

* Seattle-
  Tacoma-
  Bellevue

Chicago-
Naperville-
Joliet

Toledo

Kansas
City

New York-
Northern
New Jersey-
Long Island

Oklahoma
City

Los Angeles-
Long Beach-
Santa Ana

Philadelphia-
Camden-
Wilmington

Houston-
Baytown-
Sugar Land

Little Rock-
North Little
Rock

Lexington-
Fayette

Lowest emissions per capita

1. Honolulu
2. Los Angeles-Long Beach-Santa Ana
3. Portland-Vancouver-Beaverton, Ore.-Wash.
4. New York-Northern New Jersey-Long Island
5. Boise-Nampa, Idaho

Highest emissions per capita

100. Lexington-Fayette, Ky.
99. Indianapolis
98. Cincinnati-Middletown, Ohio-Ky.-Ind.
97. Toledo, Ohio
96. Louisville, Ky.-Ind.

Source: Metropolitan Policy Program at the Brookings Institution

THE NEW YORK TIMES

Figure 5-3. The Carbon Footprints of U.S. Cities[33]

and most importantly, the dependence on coal-fired generation that supplies electrical energy in Lexington, Cincinnati and Indianapolis. New York City has a comparatively high urban density with economic disincentives for automobile ownership (such as expensive parking) plus a highly-developed public transportation system—infrastructures that keep its carbon footprint lower.

According to Marilyn Brown at Georgia Institute of Technology, cities with the highest carbon footprints "are in the traditionally regulated states... utilities are reacting to what turns a profit for their shareholders."[35] Utilities in de-regulated states often offer their customers more creative energy conservation and alternative energy incentives and programs. The Brookings Institution study makes the following recommendations for cities that desire to reduce their carbon footprints:

- Expand public transit systems
- Support planning decisions that provide incentives for compact development
- Introduce more energy efficient freight operations and engage in regional freight system planning
- Support energy efficiency improvements
- Integrate solutions for land use, transportation and energy supply systems

Policy recommendations in the report included setting a price for carbon emissions, increasing energy research and development funding, providing greater financial support for mass transit and revising federal policies that reward states for high levels of travel and fuel use.[36]

PUBLIC AND PRIVATE SECTOR PARTNERSHIPS

Public and private sector partnerships are being formed to reduce carbon and other greenhouse gas emissions. Such partnerships occur when public entities enter into legally binding agreements with private corporations. Together, these organizations have shared goals and jointly accept the costs, risks and benefits of their initiatives, such as those associated with financially supporting development projects. Complementary resources include financial capital, human capital, political influence, technical knowledge and shared expertise between the organizations to achieve common goals.[37] Landfill gas projects provide a heuristic example.

The Lancaster County Solid Waste Management Authority (LCSWMA), which serves Lancaster County, Pennsylvania, owns a landfill that is operated by Coventa Energy. The landfill annually accepts 589,300 metric tons (580,000 long tons) of municipal waste.[38] In 2005, the LCSWMA entered into a partnership with PPL Energy Services LLP to develop landfill gas production facilities. The partnership co-developed the well fields, installed a landfill gas (LFG) pipe-conveyance system, built a road to the site and constructed a 3.2 MW power plant, using two Caterpillar 3220 internal combustion engines.[39] The LCSWMA invested

$1.5 million in this project, which became operational in February 2006.[40]

Carbon-reduction projects can be used to generate carbon exchange offsets. To initially qualify to earn carbon credits under the rules of the Chicago Climate Exchange (CCX), the Lancaster County project had to meet a set of requirements that included:[41]

- The LFG collection system had to have been installed after January 1, 1999.
- The applying organization had to retain ownership of the LFG collection system.
- The applying organization had the legal rights to any offsets, allowances, or other environmental attributes related to the LFG emissions.
- The LFG collection system had to be installed voluntarily (i.e., removal of LFG emissions must be voluntary and in the absence of any regulatory requirements from the U.S. EPA or state regulatory agency).
- The LFG generated from the site must be destroyed (e.g., by flaring, internal engine combustion, etc.).

In addition, the LCSWMA was required to become a member of the CCX. Under this affiliation, the LCSWMA had to baseline their carbon output and make a voluntary, but legally binding, commitment to reduce GHG emissions. In this case, the LCSWMA paid CCX an initiation and annual membership fee totaling $10,000, plus an offset verification fee of $4,750.[42] In October 2006, they received approval for 356 carbon financial instruments (CFIs) that represented 35,600 metric tons of CO_2 equivalents.[43] The LCSWMA estimated that the project would destroy 3,506 metric tons of methane gas, which is equivalent to 64,000 metric tons of CO_2, and yield $192,000 in revenue in 2007.[44]

Airport authorities provide yet another example of public sector and governmental partnerships. Despite emitting 189.6 billion kilograms (418 billion pounds) of CO_2 in 2007, U.S. airlines have been slow to focus on carbon reduction, except when driven by the economic necessity of reducing fuel costs.[45] "Airports have been spending hundreds of millions on terminal facilities that are esthetically pleasing but aren't designed to conserve energy" according to the Air Transportation Association, an airline industry trade group.[46] Not unlike many commercial and industrial concerns, airports require quick returns on their investments and often returns on environmental projects are "not quick enough for the airports."[47] Regardless, airports are perceived by the public to be polluters. Airport authorities often want to change this perception. Using alternative energy systems offers a highly visible solution.

Boston's Logan Airport has installed 20 wind turbine generators at the airports headquarters that will generate 100,000 kWh annually, 3% of the building's total energy requirements, at a cost of $140,000.[48] Denver International, in Colorado, installed 9,200 solar collector panels that generate 3 million kWh per year.[49] The electricity from the solar arrays supply half of the energy needed for the airport's people-mover rail system. The $15 million investment was offset by credits provided by the airport's electrical utility. In July 2008, Fresno Yosemite Airport, in California, installed 11,700 solar panels, adequate to provide 40% of total electric requirements.[50] Louisville International Airport in Kentucky replaced its fluorescent lighting systems, installed skylights in terminals and upgraded it central energy managements system.

INTERNATIONAL LOCAL GOVERNMENTAL INITIATIVES

Amersfoort, Netherlands, has created a solar-powered suburb as part of its effort to become a sustainable city. The medieval city, with its narrow streets and squares, was developed for pedestrian access and retained its character as the commercial center of the Amersfoort. This city of 130,000 is expanding to accommodate 160,000 residents by 2016.[51] The newer developments incorporate parks, canals and wetlands. In Nieuwland, a newly development suburb on the north side of the city, roughly 85% of its structures, including residences, social housing, commercial and institutional buildings have been designed with south facing roofs to optimize solar gain. Vehicular access and parking are strictly controlled. Two elementary schools and a common sports complex are creatively designed to generate soar power. At the sports complex, solar photovoltaic panels are installed on roofs and canopies. The buildings also use solar power for water heating. Photovoltaic panels generate 1.35 megawatts of electrical power for the community. Pilot demonstration homes are grid connected and generate adequate electricity for all of their needs. The environmental impact of this quarter is estimated to reduce carbon emissions by 89,000 kilograms (roughly 98 tonnes) annually.[52]

SUMMARY

Local governments have a wide range of choices that impact sustainability and the ways that they can achieve it. These choices go beyond simply changing energy consumption and energy production patterns and styles of energy demand management programs. Collectively, they are important to reducing carbon emissions. The environmental impacts of energy production can be mitigated by greater use of renewable energy, cogeneration, district heating and cooling systems and switching from coal use to natural gas. Their choices are broad and include considering the types of policies and programs that cities pursue, how they manage infrastructure, and how they purchase energy and energy-consuming equipment.

The policies and programs considered in this chapter suggest that city agendas are in flux and that many relate to the goal of reducing carbon emissions. There are regional differences in the carbon footprints of cities. Those in the western U.S. are commonly far less carbon intensive than cities in other regions of the county.

Actions by cities to include inventorying greenhouse gases, updating land use policies and building codes, encouraging use of public transit systems, supporting alternative energy, promoting green construction practices, improving recycling programs and promoting tree planning to absorb carbon dioxide. Portland, Oregon, was the first U.S. city with a global warming action plan. Local governments are developing financial incentives to help "green" their cities. Austin, Texas, plans to make its facilities more energy efficient, power them with renewable energy and intends to be carbon neutral by 2030. Cambridge, England, set a goal in 2003 of reducing its carbon emissions by 60% by 2025 and plans to achieve this by inventorying carbon emissions, directing resources to effectively lower their emissions, and implementing an emissions monitoring program. The Netherlands has created a solar-powered suburb in the city of Amersfoort.

Public-private sector partnerships provide opportunities to improve energy-efficiency and reduce the carbon intensity of governmental entities. Examples of such partnerships discussed in this chapter included city and county governments and airport authorities.

Solar Power Residences in Amersfoort, Holland

Rooftop Solar Collector Arrays on Activity Center in Amersfoort, Holland

Endnotes

1. Austin City Connection (2007, February 7). *Wynn announces Austin climate protection plan.* http://www.ci.austin.tx.us/council/mw_acpp_release.htm, accessed 6 March 2008.
2. Sedensky, M. (2006, October 25). *Mayors pick up where Washington failed on Kyoto.* http://www.iht.com/articles/2006/10/23/business/kyoto.php, accessed 27 April 2008.
3. Konch, W. (2008, August 7). Builder codes turn a "green leaf." *USA Today.* p. 3A.
4. *Ibid.*
5. One metric ton is the equivalent of 1,000 kilograms (kg), 2,679 pounds (troy), 1.1023 short tons or 0.9842 long tones.
6. Climate Change (2006, September 25). *Conversion of 1 MMT CO$_2$ to familiar units.* http://arb.ca.gov/cc/factsheets/1mmtconversion.pdf, accessed 2 January 2008.
7. *U.S. Mayor's Climate Protection Agreement.* www.seattle.gov/mayor/climate, accessed 11 June 2007.
8. *Ibid.*
9. The U.S. mayor's climate protection agreement (2005). http://www.usmayors.org/climateprotection/documents/mcpAgreement.pdf, accessed 29 April 2005.
10. Earth Policy Institute (2006, May 12). *U.S. Mayors pledge to cut greenhouse gases while Bush administration takes no action.* http://www.citymayors.com/environment/usmayors_kyoto.html, accessed 29 April 2008.
11. *Ibid.*
12. *Ibid.*
13. *Ibid.*
14. *Ibid.*
15. For more information see www.mbta.com.
16. Skoloff, B. (2007, December 28). Cities offer cash, perks to "go green." *The Courier-Journal.*
17. Johnson, E. (2008, September). Arlington County, Virginia. *Cities Go Green*, p. 15.
18. *Ibid.*
19. *Ibid.*, p. 16.
20. Skoloff, B. (2007, December 28). Cities offer cash, perks to "go green." *The Courier-Journal.*
21. Mayor's Office of Communications (2007, December 18). *Mayor Newsom unveils first-ever carbon offsets to fight global warming.* www.sfgov.org/site/mayor_index.asp?id=72509, accessed 2 May 2008.
22. *Ibid.*
23. City of Austin (2007). *Austin climate protection plan.* http://www.ci.austin.tx.us/council/downloads/mw_acpp_points.pdf, accessed 7 March 2008.
24. Muraya, N. (2008). Austin climate protection plan "possibly the most aggressive City greenhouse-gas reduction plan." *Energy Engineering.* 105 (2). p. 35.
25. *Ibid.*, p. 34.
26. Austin climate protection plan (2007, March 7). *The Austin Chronicle.* http://www.austinchronicle.com/gyrobase/Issue/story?oid=oid%3A453480, accessed 6 March 2008.
27. From CRed. *The community carbon reduction project at UNC—Chapel Hill.* http://www.ie.unc.edu/content/research/cred/cambridge.html, accessed 28 April 2008.
28. Kuck, S. (2008, May 29). The geography of America's carbon footprint. http://www.worldchanging.com/archives/008068.html, accessed 17 October 2008.

29. See Roosa, S. (2008). *The sustainable development handbook*. Chapters 7, 8 and 9. The Fairmont Press.

30. The measurement system developed by the Brookings Institute was created by three researchers: Marilyn Brown, Frank Southworth and Andrea Sarzynski.

31. Kuck, S. (2008, May 29). *The geography of America's carbon footprint.* http://www.worldchanging.com/archives/008068.html, accessed 17 October 2008.

32. *Ibid.*

33. Barringer, F. (2008, May 29). Urban areas on West Coast produce least emissions per capita, researchers find. *New York Times.* http://www.nytimes.com/2008/05/29/us/29pollute.html?ref=us, accessed 19 October 2008.

34. *Ibid.*

35. *Ibid.*

36. *Ibid.*

37. The Climate Group (2007). *Public-private partnerships: local initiatives 2007.*

38. Warner, J. (2007, November). Selling carbon LFG credits. *MSW Management.* http://www.mswmanagement.com/mw_0711_selling.html, accessed 2 March 2008.

39. *Ibid.*

40. *Ibid.*

41. *Ibid.*

42. *Ibid.*

43. *Ibid.* The commodity traded by the CCX is a contract that represents the equivalent of 100 metric tons of carbon-dioxide.

44. *Ibid.* Revenue was estimated based on a sale price of roughly $3/metric ton of CO_2.

45. Yu, R. (2008, September 17). Airports go green with eco-friendly efforts. *USA Today.* p. B2.

46. *Ibid.*

47. *Ibid.*

48. *Ibid.*, p. B1

49. *Ibid.*, p. B2

50. *Ibid.*

51. *Amersfoort, city with a heart.* http://www.amersfoort.nl/smartsite.shtml?id=51831, accessed 22 October 2008.

52. Wheeler, S. and Beatley, T. (2004). *The sustainable urban development reader.* Routledge. p. 304.

Chapter 6

Carbon Reduction Strategies for Buildings

"The National Building Museum here in Washington, D.C. was built in the 1880s. It took energy to manufacture or extract the building materials and transport them to the construction site, plus more energy to erect the building. When you add it up, the total embodied energy in the National Building Museum is equivalent to nearly 1.2 million gallons of gasoline. If the average vehicle gets about 21 miles per gallon, there is enough embodied energy in that one building to drive a car 25 million miles. If the building were demolished, all that energy would be utterly wasted."

Richard Moe, President National Trust for Historic Preservation[1]

The idea that buildings are major carbon generators is not new. Buildings contain electrical and mechanical equipment and their thermodynamic processes use energy, much of it generated from fuels containing carbon. In the U.S., buildings accounted for 72% of all electrical use and 80% of all expenditures on electricity in 2005.[2] In addition, buildings represent 39% of primary energy use (including production), 12% of potable water use, and generate 136 million tons of construction waste annually (USGBC October 2006:1-3).

The building sector accounts for roughly half of all greenhouse gas emissions and approximately 38% of CO_2 emissions in the U.S.[3] Emissions of CO_2 attributable to buildings are greater than either the industrial or transportation sectors. The residential sector alone contributes over 1.12 billion metric tons of CO_2 per year to U.S. carbon emissions.[4] As cities continue to expand, adding new structures and extending services, the need for energy resources increases, as does the potential to increase carbon emissions. There is great opportunity for architects, engineers and building design professionals to incorporate sustainable design practices and appropriate technologies in new construction and existing building improvements.

The carbon associated with the energy used in U.S. buildings alone "constitutes 8% of the current global emissions."[5] Strategies for reducing the carbon impact of buildings focus on reducing their fossil fuel requirements. One strategy is to replace windows in existing buildings. In the U.S., there are roughly 120 million windows in residential structures and 64% are single pane.[6] Newer, energy-efficient windows reduce the transmission of conditioned air by 60-65% or more. Replacing single pane windows in the U.S. with more efficient ones would result in an energy reduction equivalent to what can be produced by over a hundred 600 MW coal-fired electrical generating plants. Yet at the current replacement rate of 2.5 million window units annually, roughly fifty years will be required to replace the entire stock of single-pane windows in the country. Replacing the country's single pane windows can be accomplished at a cost of only $75-100 billion, far less than expanding electrical capacity and with much less environmental impact. Reducing electrical demand is yet another strategy that reduces the need for additional power generation facilities. The potential of this strategy is significant. It has been estimated by the McKinsey Global Institute that among the residential, commercial and industrial sectors, energy demand for buildings has the potential to be reduced by 15 quadrillion British thermal units (QBtu) or 19% of the total 92 QBtu being consumed annually.[7]

Since buildings require substantial inputs of resources to construct and maintain, what can be done to reduce their environmental impacts and energy requirements? The solutions include but are not limited to:

- Changing land development practices;
- Designing buildings with attention to improved construction standards;
- Upgrading and reusing existing structures;
- Providing more efficient buildings and higher quality equipment;
- Changing the physical arrangement and configuration of buildings;
- Carefully selecting construction materials; and
- Harvesting on-site resources.

How buildings and infrastructure are designed is important to reducing carbon emissions. This is due to variations in the carbon intensity of the materials and building components. It is also due to the variability in the types and amounts of energy required for the construction and operation of buildings. The total carbon emissions caused by buildings is increasing. Carbon emissions originating from U.S. buildings increased 47.6% from 1980 to 2005 and are projected to continue to increase through 2030.

Table 6-1. CO_2 Emissions of U.S. Buildings[8]
(10^6 metric tons of carbon)

	Fossil	Electricity	Total	Growth Rate 2005 — Year
1980	172.0	255.2	427.1	—
1990	153.7	317.2	470.9	—
2000	167.4	426.2	593.5	—
2005	**164.3**	**466.0**	**630.3**	—
2010	168.7	498.4	667.1	1.1%
2015	177.0	539.4	716.4	1.3%
2020	180.9	579.5	760.4	1.3%
2025	184.0	635.4	819.4	1.3%
2030	187.7	697.7	885.4	1.4%

In 2005, the level of CO_2 emissions from U.S. buildings, approximately 630 MMT, accounted for 39% of U.S. total carbon emissions, a quantity equal to the combined emissions of Japan, France and the United Kingdom.[9]

THE CONCEPT OF GREEN BUILDINGS

There was a time in the U.S. when most construction material was obtained locally. Indigenous raw materials were limited and included accessible timber, fieldstone, quarried rock, adobe, thatch, slate, clapboard, and cedar shakes, all of which continue to be available today. Since construction materials were costly to manufacture (most were hand-made) and troublesome to transport, most components of demolished structures were reused in some manner.

Early in U.S. history, one and two-room log houses were the norm. The central heating system was a drafty fireplace with a chimney

Table 6-2. Carbon Emissions of U.S. Buildings (2005)

2005 Buildings Energy End-Use Carbon Dioxide Emissions Splits, by Fuel Type (Million Metric Tons of Carbon Equivalent (MMTCE))

	Natural Gas	Distil.	Resid.	LPG	Oth	Petroleum Total	Coal	Elec.	Total
Space Heating	70.1	20.0	2.9	4.5	2.3	29.7	2.9	37.8	140.5
Lighting								113.7	113.7
Space Cooling	0.4							81.5	81.9
Water Heating	24.6	3.7		0.8		4.5		30.8	59.9
Refrigeration								38.8	38.8
Electronics								44.5	44.5
Cooking	6.4			0.5		0.5		14.1	21.0
Ventilation								17.6	17.6
Wet Clean	1.0							15.9	16.9
Computers								13.5	13.5
Other	4.4	0.5		4.5	0.9	5.9		39.7	50.0
Adjust to SEDS	10.3	3.6				3.6		18.0	31.9
Total	**117.2**	**27.8**	**2.9**	**10.3**	**3.2**	**44.2**	**2.9**	**465.9**	**630.2**

Source: *Building Energy Data Book* (September 2007).

constructed of local stone. When possible, design features, learned by trial and error, were added in an attempt to optimize thermal comfort. Examples included architectural features that controlled lighting and shading devices that tempered the indoor environment. Fenestration was carefully sized and oriented take advantage of natural breezes. Rainwater was often collected from roofs. Such were the humble beginnings of green construction practices.

A variety of practical energy resources were readily available and used during the early history of the U.S. Forests were cleared for agricultural purposes and the wood stocks were harvested for building materials and fuel. In order to provide heating, both wood and coal were primary fuels that were burned in fireplaces. There was little concern for the pollution that was generated. Whales were harvested—almost to extinction—for their oil to provide lighting. Buildings had no central air conditioning. Cities thrived in moderate climates and struggled in locations that experience hot and humid summers. Charleston, South Carolina, in its early history designed its residences with tall windows to take advantage

One Room Log Home in Danville, Kentucky

of sea breezes and shaded their facades with large covered porches. On warm summer nights, these porches served as social activity areas.

Buildings today are infinitely more complex—and high performance building practices can be even more so. Accepting the notion that sustainable, environmentally appropriate, and energy-efficient buildings can be labeled "green," the degree of "greenness" is subject to multiple interpretations. The process of determining which attributes of a structure can be considered "green" or "not green" can be viewed as subjective—yet there are attributes of green construction practices that are universally accepted.

A green design solution appropriate for one locale may be inappropriate for others, due to variations in climate or geography. Gauging the degree of "greenness" involves understanding and applying standards that are often formative and evolving. While it is implied that green buildings are an improvement over current construction practices, comparisons are often unclear and confusing. Determining what sort of changes in construction might lead to greener, more sustainable buildings can be perplexing. While markets adjust to provide environmentally friendly

Multifamily Residences in Charleston, South Carolina

materials, components and products for use in the construction of greener buildings, green standards are evolutionary and continually changing.

In Pennsylvania, the Governor's Green Government Council defines a green building as "one whose construction and lifetime of operation assures the healthiest possible environment while representing the most efficient and least disruptive use of land, water, energy and resources."[10] Attributes of green construction practices include preserving and restoring habitat that is vital for life and becoming "a net producer and exporter of resources, materials, energy and water rather than being a net consumer."[11] Another attribute of green construction is reducing carbon intensity. Lower carbon emissions equate to less environmental impact.

There are qualities of structures, such as reduced environmental impact, improved indoor air quality and comparatively lower energy usage, which are widely accepted as evidence of green construction practices. For example, using recycled materials that originate from a previous use in the consumer market, or using post-industrial content that would otherwise be diverted to landfills, are both widely accepted green construction practices.

Grumman (2003:4) describes a green building by saying that "...a green building is one that achieves high performance over the full life cycle." Performance can be defined in various ways. "High performance" can be interpreted widely, and at times in "highly" subjective ways. High performance can be considered to be a "segment of the green building industry that focuses on how efficiently a building uses energy."[12] However, the idea has been considered in much broader terms. In an attempt to clarify, Grumman further identifies a number of attributes of green buildings:

- Minimal consumption—due to reduction of need and more efficient utilization—of nonrenewable natural resources, energy resources, land, water, and other materials as well.

- Minimal atmospheric emissions yielding less negative environmental impacts, especially those related to greenhouse gases, global warming, particulates, or acid rain.

- Minimal discharge of harmful liquid effluents and solid wastes, including those resulting from the ultimate demolition of the building itself at the end of its useful life.

- Minimal negative impacts on site ecosystems.

- Maximum quality of indoor environment, including air quality, thermal regime, illumination, acoustics, noise and visual aspects.

High performance implies added value. Energy-consuming building systems that are used in high performance buildings require specialized design expertise and non-traditional system components. In fact, adding green building features is likely to increase initial construction costs by 1% to 5%. However, lower operating expenses and reduced life cycle costs typically offset increases in initial construction costs.

GREEN BUILDING CONSTRUCTION PRACTICES

While attributes of green construction practices are readily identified, developing construction standards to achieve such goals is another matter. Factors that contribute to building differentiation include the desires of the owners, the skills and creativity of their design teams, site locations, local planning and construction standards, and a host of other conditions. What may be a green solution for one building might be inappropriate if applied to another. Sustainable buildings have high performance and green attributes. Sustainable design, energy efficiency and the carbon footprint of buildings are inherently linked. How so? Energy efficient structures use less carbon-based fuels and result in buildings with smaller carbon footprints. As a result, they are more sustainable.

The American Institute of Architects' (AIA) Sustainable Architectural Practice Position Statement advocates sustainable design that includes reduction in the use of non-renewable energy resources and promotes "resource conservation to achieve a minimum 50% reduction from the current level of consumption of fossil fuels."[13] The AIA's Committee on the Environment's Measures of Sustainable Design provides a number of tenets that identify and qualify these practices.

While the qualities of green construction practices can be identified, there are shades of green in green building construction practices.[14] Some qualities focus on exterior features and others on the types of green materials that are used. These might include enhanced insulation, high-performance windows, shading devices, or reflective roofing systems. The 2009 International Energy Conservation Code (IECC), for example, requires that energy efficient design be used in construction and provides effective methodologies. Its focus is on the design of energy-efficient building exteriors, mechanical systems, lighting systems and internal power systems. The IECC is being included in building codes across the U.S. and in many other countries. The American Society of Heating, Ventilation and Air-Conditioning Engineers (ASHRAE) published its Standard 90.1 "Energy Standard for Buildings Except Low-Rise Residential Buildings," a document that deals with the energy-efficiency of buildings and HVAC components. Standards often focus on improved air quality and better ventilation systems. Green construction opportunities are also identified in the *ASHRAE Green Guide*.

GREEN CONSTRUCTION METHODOLOGIES

Material and product recycling has a long history in green construction and has its roots in byproduct recycling. During World War II, strategic materials, such as steel and aluminum, were recycled and reused to manufacture military equipment. After the end of the war, recycling programs fell into decline. Beginning anew in the 1970s, metals such as aluminum, copper and steel began to be recycled. By the

Table 6-1. AIA Measures of Sustainable Design

Measure 1: Design & Innovation
Sustainable design is an inherent aspect of design excellence. Projects should express sustainable design concepts and intentions, and take advantage of innovative programming opportunities.

Measure 2: Regional/Community Design
Sustainable design values the unique cultural and natural character of a given region.

Measure 3: Land Use & Site Ecology
Sustainable design protects and benefits ecosystems, watersheds, and wildlife habitat in the presence of human development.

Measure 4: Bioclimatic Design
Sustainable design conserves resources and maximizes comfort through design adaptations to site-specific and regional climate conditions.

Measure 5: Light & Air
Sustainable design creates comfortable interior environments that provide daylight, views, and fresh air.

Measure 6: Water Cycle
Sustainable design conserves water and protects and improves water quality.

Measure 7: Energy Flows & Energy Future
Sustainable design conserves energy and resources and reduces the carbon footprint while improving building performance and comfort. Sustainable design anticipates future energy sources and needs.

Measure 8: Materials & Construction
Sustainable design includes the informed selection of materials and products to reduce product-cycle environmental impacts, improve performance, and optimize occupant health and comfort.

Measure 9: Long Life, Loose Fit
Sustainable design seeks to enhance and increase ecological, social, and economic values over time.

Measure 10: Collective Wisdom and Feedback Loops
Sustainable design strategies and best practices evolve over time through documented performance and shared knowledge of lessons learned.

Source: American Institute of Architects, http://www.aia.org/SiteObjects/files/Description%20of%20Measures%20of%20Sustainable%20Design.pdf, accessed 5 March 2008.

1980s, construction site wastes, such as steel frame windows and their glass panels, were being recycled rather than being sent to landfills. By the late 1990s landfill space became more costly and material recycling was expanded to include wood, brick, tile and other construction components. In addition, once-flared natural gas from landfills began to be seen as a potential energy resource rather than a waste by-product. Recently increases in the costs of fuels such as oil and natural gas have created conditions which make landfill gas an economically viable fuel.

Waste Materials from a Construction Site, Pioneer Village, Kentucky

Clayton (2003:486) defines a "sustainable" building philosophy as "to design, build and consume materials in a manner that minimizes the depletion of natural resources and optimizes the efficiency of consumption." Green construction practices have several categorical commonalities:

- Green buildings are designed to reduce energy usage while optimizing the quality of indoor air. These buildings achieve energy reductions by using more insulation and improved fenestration, by optimizing the energy usage of mechanical and electrical building subsystems, and by the use of alternative energy.

- There is an emphasis on reducing the costs of energy used to transport material to the construction sites. One means of achieving reductions in transportation costs is to use materials that have been locally manufactured. This provides the added benefit of supporting local employment and industries.

- There is a focus on using recycled construction materials (such as reusing lumber from demolished structures) or materials made from recycled products (such as decking materials that use recycled plastics). The goal is to reduce the amount of virgin material required in construction.

- There is a preference for materials that are non-synthetic, meaning that they are produced from natural components such as stone or wood, etc. This often reduces the number of steps required for product manufacture and may also reduce the use of non-renewable resources. Extracted metals such as aluminum and copper may also be preferred as they can be more easily reused once the building's life cycle is completed.[15]

- There is a mandate of green construction to avoid the use of materials that either in their process, manufacture or application, are known to have environmentally deleterious effects or adversely impact heath. Examples include lead in paint or piping, mercury thermostats, and solvents or coatings that may outgas fumes and carcinogens.

- In an effort to reduce water from municipal sources and sewage treatment requirements, green buildings are often designed to harvest rainwater by using collection systems. Filtered rainwater may be used for irrigation, toilets or other non-potable requirements. In addition, green buildings use technologies, such as flow restriction devices, to reduce the water requirements of the building's occupants. Use of potable water for site irrigation is minimized.

- The quantities of construction wastes are reduced. This occurs by strategically reducing wastes generated during construction and by reusing scrap materials whenever possible. The goal is to reduce the amount of material required for construction and to reduce the quantities of scrap material that must be trucked to a landfill.

New industries and entire product lines have emerged in an effort to provide construction materials that meet green building standards. There are numerous creative examples. Beaulieu Commercial has brought to market a carpet tile backing made from 85% post-consumer recycled content using recycled plastic bottles and glass, yielding carpet backing that is 50% stronger than conventional carpet backing. The Mohawk Group manufactures carpet cores using a recycled plastic material rather than wood. This change saves the equivalent of 68,000 trees

Shading Devices for an Office Building in Kansas City, Missouri

annually. There are products on the market (e.g., ProAsh), made with fly ash—a once wasted by-product that has been recycled into concrete. Armstrong now offers to pick up old acoustical tile from renovation projects and will deliver them to one of their manufacturing sites to be recycled into new ceiling tiles. Polyvinylchloride (PVC) products became available to satisfy piping needs in domestic applications for drainage systems rather than copper, lowering the weight of products and their components. Eliminating the need to sweat pipes lowers labor costs and reduces the installation times. While PVC has its own associated environmental issues, it is now being used for exterior trim due to its durability. Many wood products are certified and labeled if they have used environmentally appropriate growing and harvesting techniques in their production. Pervious paving systems, that allow vegetation to grow and reduce the heat absorption from paving, are also available. The list of available green technologies and products seems endless. Triple pane, low "e" window glazing, solar-voltaic roof shingles, LED lighting systems, waterless toilets and solar powered exterior lighting systems are among the green building products that are being used in building systems. There are solar collector systems that simultaneously generate electricity and provide heat for domestic hot water systems.

Green construction has provided opportunities to introduce new product lines and the movement to green construction practices offers ready markets for the products. Hydrotech manufactures a green roof system that provides a balance between water drainage and soil retention thus allowing roof gardens to flourish. As green construction practices evolve and new technologies enter the marketplace, there will be more environmentally friendly building products available in the future.

BUILDING PRESERVATION

Buildings and building materials can often be recycled and reused. Older structures can be updated to serve new purposes, recycling their existing infrastructure. Retaining existing structures by improving and reviving them sustains communities, maintains infrastructure and often

Restored Mixed-use Buildings on the Green in Litchfield, CT

Quincy Market in Boston, Massachusetts

reduces energy use. There are numerous cities that have restored areas of their city centers to create vibrant urban spaces. U.S. examples include the downtown areas of Arlington, Virginia, and Boston, Massachusetts. Small towns such as Bardswotn, KY, and Litchfield, CT, have restored their downtowns and developed mixed-use commercial centers.

This impacts energy use, since recycling building materials and building components is typically less energy intensive than manufacturing new materials. When a new building composed of 40% recycled materials is designed to green building standards, it still requires roughly 65 years to recover the energy that is normally lost when an existing structure is demolished.[16] According to Richard Moe, the President National Trust for Historic Preservation, "We have always regarded preservation as a sustainable activity because it is all about recycling resources... we are giving it more emphasis now because of public concern about global warming CO_2 emissions, and energy conservation."[17]

GREEN BUILDING RATING SYSTEMS

While green building components are available, incorporating them into green buildings requires forethought, engineering, and creative design. Green building standards have been developed by both private and governmental organizations bent on finding ways to assess green construction practices. Comparing the degree of "greenness" from one building to the next is difficult. One solution is the use of categorical rating systems in an effort to reduce subjectivity. The development of green building attributes or standards by private organizations requires that decisions are to be based on stakeholder consensus. These stakeholders are often from widely diverse industries and geographic locations.

Developing a rating system for green buildings is both difficult and challenging. According to Boucher, "the value of a sustainable rating system is to condition the marketplace to balance environmental guiding principles and issues, provide a common basis to communicate performance, and to ask the right questions at the start of a project" (Boucher 2004). Rating systems for sustainable buildings began to emerge in the 1990s.

The most publicized of these rating systems first appeared in the U.K., Canada, and the U.S. In the U.K., the Building Research Establishment Environmental Assessment Method (BREEAM) was initiated in 1990. BREEAM™ certificates are awarded to developers based on an assessment of performance in regard to climate change, use of resources, impacts on human beings, ecological impact and construction management. Credits are assigned based on these and other factors. Overall ratings are assessed according to grades that range from pass to excellent (URS Europe 2005).

The international initiative for a Sustainable Built Environment (iiSBE), based in Ottawa, Canada, has a Green Building Challenge program that is now used by more than 15 participating countries. This collaborative venture provides an information exchange for sustainable building initiatives and has developed "environmental performance assessment systems for buildings" (iiSBE 2005). The iiSBE has created one of the more widely used international assessment systems for green buildings.

The U.S. Green Building Council (USGBC), an independent non-profit organization, grew from just over 200 members in 1999 to 13,000 members by 2008.[18] The core purpose of the USGBC "is to transform the way buildings and communities are designed, built and operated, enabling an environmentally friendly, socially responsible, healthy and prosperous environment that improves the quality of life" (USGBC 2006:1). The program developed by the USGBC is called LEED™, for Leadership in Energy and Environmental Design.

Prior to the efforts of organizations like the USGBC (established in 1995), the concept of what constituted a "green building" in the U.S.

lacked a credible set of standards. The USGBC's Green Building Rating System has a goal of applying standards and definitions which link the idea of high performance buildings to green construction practices.

Sustainable technologies are firmly established within the LEED project development process. LEED loosely defines green structures as those that are "healthier, more environmentally responsible and more profitable" (USGBC 2004).

The LEED Green Building Rating System is a consensus-developed and reviewed standard, which allows voluntary participation by diverse groups of stakeholders interested in its application. The LEED system is actually a set of rating systems for various types of green construction projects. Projects are ranked as part of the labeling process. LEED rating systems are available for new construction, existing buildings, commercial interiors, core and shell development projects, schools, residences, and neighborhood development.[19]

The LEED rating systems evolved to help "fulfill the building industry's vision for its own transformation to green building" (USGBC 2004). The first dozen pilot projects using the rating system were certified in 2000. By 2006 there were 27,200 LEED Accredited Professionals, 3,539 LEED registered projects in 12 countries, and 484 projects that have completed the certification process (USGBC, October 2006). Additional LEED rating systems are being developed and separate rating programs for the construction of healthcare and retail facilities. The LEED rating system is poised to become the new international standard for green buildings. In March of 2009, a compact between BREEM, LEED and the UK Green Building Council was concluded "to map and develop common metrics to measure emissions of CO_2 equivalents from new homes and buildings."[20] Their plan is to ultimately develop measurable and verifiable ways to reduce the greenhouse gas emissions generated by buildings.

ENERGY STAR BUILDINGS

In an effort to provide information to improve the energy efficiency of buildings in the U.S., agencies of the central government co-sponsored the development of the Energy Star™ program. This program provides "technical information and tools that organizations and consumers need to choose energy-efficient solutions and best management practices" (USEPA 2003). The type of technical information available includes information about new building designs, green buildings, energy efficiency, networking opportunities, plus web-accessible tools and other resources. Energy Star offers opportunities for organizations and governments to become partners in the program. The program offers guidelines to assist organizations in improving energy and financial performance in an effort to distinguish their partners as environmental leaders. The multi-step process involves making a commitment, assessing performance, setting goals, creating an action plan, implementing the action plan, evaluating progress and recognizing achievements.[21]

The Energy Star building energy performance rating system has been used for over 10,000 buildings. Energy Star does not claim to be a green building ranking system but rather a comparative assessment system that focuses on energy performance, an important component of green building technologies. The rating system is based not on energy costs but on source energy consumption. In the view of the Energy Star program, "the use of source energy is the most equitable way to compare building energy performance, and also correlates best with environmental impact and energy cost."[22] Using the Energy Star rating system, buildings are rated on their energy performance on a scale of 1 to 100 when compared to similar structures constructed in environments that experience similar weather conditions. Buildings that achieve a rating of 75 or greater qualify for the prized Energy Star label.[23]

ZERO-ENERGY BUILDINGS

Zero-energy buildings are those that are designed to be independent from utility power grids or connected primarily for the purpose of providing marginal quantities of supplemental power. They strive to use alternative energy systems to generate their power from on-site systems to achieve the capability of being carbon neutral. According to Jan Dokkum of UTC Power, "Buildings of tomorrow should be self-sufficient in energy and have carbon neutral emissions."[24] This can be accomplished by "incorporating renewable energy sources into a building's design, optimizing energy efficiency of support systems, and taking advantage of geographic and culturally acceptable building practices."[25] Zero-energy measures have been defined as installations that use passive building design practices (such as natural cooling), active solar systems (such as photovoltaic), biomass-fuels and energy monitoring (Newman 2008:135).

While there are commonalities, some buildings called zero-carbon buildings focus solely on the carbon content of the energy used for operation of the building while others are designed to minimize energy usage. These structures can be designed to minimize their internal energy requirements by using ultra-efficient building components and systems while maximizing the production of energy from on-site mechanical and electrical systems. These buildings are often called zero-energy buildings (ZEBs). Designed to require no energy for their operation that is not generated on site, they must use alternative energy sources. In the London Borough of Sutton, an environment-friendly mixed-use housing development named the Beddington Zero Energy Development (BedZED) includes 82 residences, 17 apartments plus work spaces. To achieve zero energy use, the project uses solar energy, enhanced natural ventilation, tree waste fuels, and a cogeneration plant. In addition, the development incorporates low impact materials, a waste recycling system and a car-sharing program.

It has been estimated that the residential sector accounts for 17% of U.S. green house gas emissions or 3.5% of total global emissions.[26] This amount equates to over 1.12 billion metric tons of CO_2 annually.[27] Reducing the energy use of residences reduces carbon emissions.

Zero energy homes (ZEH) are often connected to the electric grid and supplement any grid power when their energy producing systems are inadequate to satisfy internal requirements. However, when energy-generating systems produce a surplus, electrical power is supplied to the grid. A ZEH is designed to produce enough renewable power from on-site sources to offset any non-renewable power purchased from the utility so that the result is a net-zero annual energy bill.[28] Renewable sources of energy for these residences often includes using solar and wind power for electrical generation and geothermal energy for heating and cooling requirements. While zero-energy homes are rare there are examples. A zero-energy home in Panama City, Florida, recently received a LEED Platinum certification. The features of the residence include solar panels, a geothermal system for heating and air conditioning, foam insulation, a high-tech air filtration system, high efficiency hot water heater plus Energy Star appliances.[29]

Regardless, designing and constructing workable zero energy structures can be challenging. According to Dave Hewitt, the Executive Director of the New Building Institute, "While net zero is a worthy goal, these buildings are not easily attainable with today's technologies... but that doesn't mean that progress can't be made."[30] He believes that high performance structures are a continuum from advanced practices through to buildings that harvest on-site renewable energy, evolving to net-zero construction technologies. Such buildings would require so little energy that the balance of the energy requirements can be provided with non-carbon energy sources. Beyond these are buildings that produce more energy than they consume.

NET POSITIVE ENERGY CONSTRUCTION

Net positive energy (NPE) construction involves buildings that actually generate energy. While these buildings may require external sources of energy, energy generating systems in place typically produce far more energy than they consume. How is this possible? In theory, NPE construction involves reducing energy use to a minimum while maximizing the use of non-carbon emitting energy resources available on the building site. For residential construction this requires a number of features that must be considered in the building systems:[31]

- Maximizing the potential of sustainable site resources.
- Climate responsive landscape and architectural design.
- An exceptionally efficient thermal envelope.
- Reducing solar loads by "tuning" the exterior features of the building (e.g., exterior colors and surfaces).
- The installation of highly efficient heating, cooling, hot water and lighting systems.
- Minimizing water consumption by using water collection systems.
- Reducing on-site consumption of fossil fuels.
- Reducing electrical consumption and demand to micro-load levels by using extremely efficient appliances and scheduling their use.
- Energy management systems that assist in monitoring and controlling electrical loads.

Comparative modeling studies indicate that adding these features to residences will increase their costs. To achieve the goals of a NPE residence, the initial premium for new homes has been estimated to be 16.5-18.3% of the total construction costs.[32] However, when construction is financed the increases in mortgage costs (including both principle and interest expense) incurred by the homebuyer can often be entirely offset by the decrease in utility costs over the life of the residence. Unlike utility expenses, interest expenses on mortgages are tax deductible in the U.S. Utility expenses have been increasing at a pace greater than the general rates of inflation, increasing the risks to the homeowner. These accounting and economic nuances provide hidden financial incentives to construct energy efficient homes which are often overlooked.

Commercial NPE projects are on the drawing boards. The new headquarters for Masdar in Abu Dhabi is being designed by a U.S. architecture firm and may be the world's first zero-carbon, zero-waste building that is fully powered by renewable energy.[33] The headquarters will be a mixed-use "positive energy" building that actually produces more energy than it consumes.[34] The design incorporates features that virtually eliminate carbon emissions and reduce liquid and solid wastes. Plans to achieve this include installing a building integrated solar photovoltaic array, providing solar thermal cooling and dehumidification systems, and generating power for the building's assembly. In addition to providing offices for Masdar, the building will accommodate private residences and other businesses. According to Adrian Smith, partner in the Architecture firm of AS+GG, "As a positive energy complex, the project will have a far-reaching influence on the buildings of tomorrow."[35]

CARBON NEUTRAL CONSTRUCTION

Recently, the idea that new buildings can be designed to be carbon neutral has evolved. Some believe that "buildings that use no energy from external power grids are carbon neutral."[36] While more applicable for new construction, a truly carbon neutral building is complex and must be designed so that building components are also carbon neutral. This is a very difficult undertaking. Regardless, high performance buildings can be designed to focus on carbon impact and it is theoretically possible to design them with features that allow some to be classified as "zero-carbon," meaning that they have a net zero carbon emissions impact.

In order to prove that the building is carbon neutral, the total carbon emissions content needs to be calculated. This is a difficult task. There are four primary stages of a building's life cycle that need to be considered:

1) Site Disruption—Carbon impact of changes to landscaping and preparing a site for construction.

2) Construction—Estimating the carbon content and impact of the materials, labor and activities associated with construction of the building.

3) Operating Life Cycle—Calculating the carbon impact of the building during its operating life including energy and water consumption.

4) Disposal—Carbon impact of the disposition of the building and materials.

For new construction, carbon neutral buildings must account for carbon emissions associated with the transportation of materials to the building site. Often this is not the case and accounting for the carbon content of multiple, often-interrelated sources can be mathematically complex. Given these parameters, designing a carbon neutral building is a challenging undertaking. A truly carbon neutral building may be impossible to construct. However, the remainder of the carbon can be resolved by counterbalances and offsetting carbon emissions. To achieve this, owners may choose to obtain credits from green initiatives such as energy efficiency or renewable energy.

The Aldo Leopold Foundation Building in Wisconsin is an example of an operationally carbon-neutral building. This building sits on a 32-hectare (80 acre) site of worn-out farmland and achieved a LEED Platinum certification. Architectural features of the Aldo Leopold Foundation Building include enhanced insulation, passive design that provides daylighting and heating during winter and shading during summer, cross ventilation, and operable windows.[37] The 1,114 m² (12,000 ft²) building produces 15% more energy than it consumes by using a 196 panel, 40 kW solar electric generation system.[38] To further reduce energy consumption, the radiant heating and cooling system is installed within the concrete floors and uses geothermal energy. The unique heating and ventilation system has "earth tubes" installed underground to precondition air before air is admitted into the building. The effectiveness of this concept was first proved in a residence near Princeton, Kentucky, designed by Dr. Roosa in the late 1970s, which used an underground "cool tube" system for heating, air conditioning and dehumidification.

SUMMARY

The construction, operation, and maintenance of buildings have proven to be quite significant, in terms of energy use and carbon emissions, throughout human history. The amount of CO_2e attributable to buildings surpasses either the industrial or transportation sector. Conspicuously, buildings are complex structures with numerous inputs and outputs that affect the surrounding environment. Buildings are also an integral part of human life; their importance and use spans from a familial, communal, municipal, regional, national, to an international level. As economic development increases and cities expand, so does the threat of carbon emissions. Consequently, there is great opportunity for architects, engineers and building design professionals to incorporate sustainable design practices and appropriate technologies in new construction and existing building improvements.

Many factors must be considered when designing and constructing green buildings. Examples include the carbon intensity of the materials and building components, as well as the types and amounts of energy required. How a corporation or individual disposes of wastes and uses energy, is critical. Other concerns include the relevant climate and geography of the structure in question. Consequently, buildings today are infinitely more complex and green designs for one locale are likely inappropriate for another.

Determining what sort of changes in construction might lead to greener buildings can be a perplexing task, which is why creativity is a prominent undertone in this chapter. New industries and entire product lines have emerged in an effort to provide construction materials that meet green building standards. Indeed, in response to the growing demand, research and comprehension of sustainable development, the building and construction industry has undergone major adjustments. Creativity and innovation are needed for green buildings. Cutting-edge approaches represent an enormous challenge to those who make the commitment to environment-friendly design principles.

In response, green building standards have been developed by both private and governmental organizations bent on finding ways to assess and guide green construction practices. Comparing the degree of "greenness" from one building to the next is difficult. Initially, the concept of a "green building" lacked a credible set of standards and was very subjective. However, the inception of the USGBC's Green Building Rating

System, otherwise known as LEED, in addition to the U.S. government's Energy Star program, has redirected and redefined the concept. While involvement in sustainable development and the related rating systems remains voluntary, there has been a dramatic and impressive increase in participation. The propagation of ZEBs and ZEHs, reflect both private sector and individual commitments to green building practice and thwarting global climate change. This chapter provides a glimpse of the momentum behind the green buildings movement.

Endnotes

1. Young, D. (2008, January). Building on what we've built. *Preservation*. p. 7.
2. U.S. Department of Energy (2007, September). 2007 Buildings energy data book. p. 1-3. http://www.btscoredatabook.net/docs/2007-bedb-0921.pdf, accessed 20 January 2008.
3. Johnson, L. (2007, May 7). *Building design leaders collaborating on carbon neutral buildings by 2030*. http://blog.fastcompany.com/archives/2007/05/07/building_design_leaders_collaborating_on_carbonneutral_buildings_by_2030.html, accessed 20 April 2008.
4. Howard, B. (2008, March). *Net positive energy homes: maximizing performance using advanced software systems*. Conference paper presented at Globalcon, Austin, Texas. p. 1.
5. Koomey, J., Weber, C., Atkinson, C., and Nicholls, A. (2001). Addressing energy-related challenges for the U.S. buildings sector: results from the clean energy futures study. *Energy Policy*. 29. p. 1209. Elsevier. This study provides an analysis of building sector energy use in 1997 and projects usage, carbon and dollar savings potential for the following 20 years based on two scenarios.
6. Matis, C. (2008, September). *Discussion of recent changes to ASHRAE 90.1*. NASEO Conference. Overland Park, KS.
7. Governor's Office of Energy Policy (2007, April 17). Possible and profitable: energy efficiency investments in the building sector. *Kentucky Energy Watch*. 8 (16).
8. U.S. Department of Energy (2007). *Buildings energy data kook*. Chapter 3.1, Carbon dioxide emissions for U.S. buildings. http://www.btscoredatabook.net/?id=view_book&c=3, accessed 20 January 2008.
9. U.S. Department of Energy (2007, September). *2007 Buildings energy data book. p. V.* http://www.btscoredatabook.net/docs/2007-bedb-0921.pdf, accessed 20 January 2008.
10. Governor's Green Government Council. What is a green building? *Building Green in Pennsylvania*. http://www.gggc.state.pa.us/gggc/lib/gggc/documents/whatis041202.pdf, accessed 18 January 2007.
11. *Ibid.*
12. Herbert, S. (2008, September). The spectrum of high performance buildings. *Cities Go Green*. p. 21.
13. American Institute of Architects. *Sustainable architectural practice position statement*. http://www.aia.org/SiteObjects/files/sustain_ps.pdf, accessed 5 March 2008.
14. The *ASHRAE green guide* is a primer for green construction practices.
15. Lincoln Hall, a LEED building at Berea College in Berea, Kentucky, actually uses copper downspouts.
16. Peirce, N. (2008, May 6). A 'green' Rx to save carbon: city density plus transit.

The Courier-Journal. p. A6.

17. Young, D. (2008, January). Building on what we've built. *Preservation.* p. 6.
18. U.S. Green Building Council. http://www.usgbc.org/DisplayPage. aspx?CMSPageID=1773, accessed 5 April 2008.
19. For a listing and explanation of the various rating systems developed for LEED, see the U.S. Green Building Council rating systems website. www.usgbc.org/ Display/Page.aspx?CMSPageID=222, accessed 2 January 2007.
20. UK Green Building Council (2009, March 3). *Common language for carbon in sight: leading rating tool providers to sign MOU.* http://www.usgbc.org/Docs/News/ MOU0309.pdf, accessed 3 March 2009.
21. U.S. EPA. *Superior energy management creates environmental leaders,* Guidelines for energy management overview. www.energystar.gov/index.cfm?c=guidelines. guidelines_index, accessed 3 January 2007.
22. U.S. EPA. *Be a leader—change our environment for the better.* Portfolio manager overview. www.energystar.gov/index.cfm?c=spp_res.pt_neprs_learn, accessed 3 January 2007.
23. U.S. EPA. *Portfolio manager overview.* www.energystar.gov/index.cfm?c=evaluate_ performance.bus_portfolimanager, accessed 3 January 2007.
24. Environment News Service (2009, March 29). *Buildings of the future—energy self-sufficient, carbon neutral.* www.ens-newswire.com/ens/mar2006/2006-03-29-03.asp, accessed 4 May 2008.
25. *Ibid.*
26. Howard, B. (2008, March). *Net positive energy homes: maximizing performance using advanced software systems.* Conference paper presented at Globalcon, Austin, Texas. p. 1.
27. *Ibid.*
28. Tool Base Services. *The zero energy homes project.* http://www.toolbase.org/Tool-baseResources/level4CaseStudies.aspx?ContentDetailID=2469&BucketID=2&Categ oryID=58, accessed 5 April 2008.
29. Bonts, M. (2008, March 26). *Florida's first zero-energy, LEED Platinum certified home unveiled in Panama City.* http://www.usgbc.org/Docs/News/Stalwart%20 LEED%20Release_AK.pdf, accessed 5 April 2008.
30. Hobert, S. (2008, September). The spectrum of high performance buildings. *Cities Go Green.* p. 28.
31. Howard, B. (2008, March). *Net positive energy homes: maximizing performance using advanced software systems.* Conference paper presented at Globalcon, Austin, Texas. p. 2.
32. *Ibid.,* p. 4-7.
33. Masdar. *Masdar headquarters to be located in world's first 'positive energy' mixed-use building.* http://www.prnewswire.com/cgi-bin/stories.pl?ACCT=104&STORY=/ www/story/02-22-2008/0004760606&EDATE, accessed 9 April 2008.
34. *Ibid.*
35. *Ibid.*
36. Environment News Service (2006, March 29). *Buildings of the future energy self-sufficient, carbon neutral.* http://www.ens-newswire.com/ens/mar2006/2006-03-29-03.asp, accessed 18 April 2008.
37. Chapa, J. (2007, November 6). *First LEED Platinum carbon neutral building.* http:// www.inhabitat.com/2007/11/08/first-leed-platinum-carbon-neutral-building/, accessed 1 April 2008.
38. *Ibid.*

Chapter 7

Technologies that Reduce CO₂ Impact

"We choose to go to the moon in this decade and do the other things, not because they are easy, but because they are hard, because that goal will serve to organize and measure the best of our energies and skills..."

President John F. Kennedy (1962)[1]

"Leaders at all levels are beginning to understand that our nation and the world cannot address climate change without cost-effective Carbon Capture and Storage (CCS) technology."

Michael J. Mudd, chief executive officer of FutureGen Alliance (July 2006)[2]

THE POTENTIAL IMPACT OF ALTERNATIVE ENERGY

Alternative and renewable energy refer to electricity or heat generated from renewable sources (including wind, solar, geothermal, biomass, landfill gas and low impact hydro) as well as energy substitutes and various conservation methods. Renewable energy is more environment-friendly than traditional electricity generation. Unlike fossil fuels, alternative and renewable energy emit little or no air pollution. They also do not generate radioactive wastes. Most importantly, they are replenished by naturally occurring processes. Investments in renewable energy have been increasing. Worldwide, renewable energy investments increased to $148 billion in 2007, growing 60% over 2006.[3]

In contrast to green power or electricity generated from renewable sources, "brown power" is electricity generated from environmentally hostile technology. The vast majority of electricity in the United States comes from coal (more than 50%), nuclear (nearly 20%), natural gas (roughly 17%), and large hydroelectric dams (about 7%). Only 1.6% of the U.S. energy is obtained through non-hydro renewable sources.[4] The remaining amounts of energy are from small, more portable fossil fuel sources. The brown power generators are sources of air pollution in the United States, contributing to smog and acid rain. They are also the greatest single contributor of GHGs, including CO_2 and nitrogen oxide.

Worldwide, a 20% to 25% increase in renewable energy production by 2020 should cause a considerable reduction in CO_2 emissions and the consequential environmental impacts, especially if the use of fossil fuels is stabilized or reduced. Innovative alternate energy technologies would also decrease the reliance on imported oil and fossil fuels. The use of alternative forms of energy, including bio-fuels, wind and solar power, is increasing in the U.S. Wind and solar energy output increased roughly 45% from 2006 to 2007. In addition, ethanol production by June of 2008 had increased to 633,000 barrels daily, an increase of 43%.[5]

ENERGY SYSTEMS AND THEIR IMPORTANCE
IN REDUCING ATMOSPHERIC CARBON

According to the Stern Review Report (2006) entitled, "The Economics of Climate Change," vigorous action is needed to halt and reverse the growth in GHG emissions. This report calls for a 75% reduction of total GHG emissions by 2050. This figure is consistent with a maximum of 450 ppm of CO_2e, which the report identifies as the upper limit of CO_2e before more drastic or disastrous environmental impacts begin to occur. One way of minimizing emissions is to reduce the demand for GHG-intensive goods and services that consume energy. There is substantial, technical potential for energy efficiency improvements to reduce emissions and costs. Over the past century, efficiency in energy supply has improved substantially in industrialized countries. There have also been impressive gains in the efficiency with which energy is utilized for heating, lighting, refrigeration and powering of industry and transportation. The successful and innovative techniques of the 20th century include the use of compact fluorescent light bulbs (CFLs), the substitution of natural gas for coal-to-heat processes and electricity generation, the installation of double-glazed windows, the implementation of natural day-lighting in building construction, the improvement of wall and cavity insulation, the use of heat pumps, etc.

The possibilities for further gains in energy efficiency are far from exhausted. Options for low-emission energy technologies are developing rapidly, though many remain more expensive than conventional technologies. Some of the attractive options include: on and off-shore wind energy, wave and tidal energy, solar energy, carbon capture and storage, hydrogen fuel cells, biomass/bio-energy, combined heat and power (CHP) or cogeneration, small hydro power, nuclear energy, hybrid or electric vehicles, etc. Regardless, there are social and economic barriers to implementing efficiency improvements. These include a lack of understanding of the technologies and their applicability, restricted access to capital for improvements, and reluctance to change.

ALTERNATIVE ENERGY AS A
CARBON REDUCTION SOLUTION

In conjunction with efforts towards energy efficiency and associated conventional technologies, renewable energy resources must be developed to effectively reduce GHG and CO_2 emissions in a sustainable manner. Briefly, these alternatives forms of energy are:

- Solar—Converting energy from the sun into electricity using photovoltaic (PV) panels, and domestic hot water with solar thermal collectors. Solar energy has a share of more than 99.9% of all the energy converted on Earth.[6] The solar radiation incident on the earth is weakened within the atmosphere and partially converted into other forms like wind and hydro power. In the U.S., the greatest use of solar water heating is to heat swimming pools.

- Wind—Harnessing the power of the wind using turbines. Wind power is currently the world's fastest growing and most cost-effective renewable energy technology. Of the total solar radiation incident on the outer layer of the atmosphere, approximately 2.5% is utilized for atmospheric movement. The energy contained in the moving air masses, which can be converted into mechanical and electrical energy by windmills, is a secondary form of solar energy.[7] Wind is generated as equalizing currents, essentially as a result of

Solar Collector System for Swimming Pools, Fort Wright, Kentucky

varying temperature levels on the surface of the earth, by which differences in air pressure have been created. The air masses then flow from higher pressure areas to lower pressure areas.

- Geothermal—Use of steam that lies below the earth's surface to generate electricity. The energy flowing from the interior of the earth to its surface is fed by three different sources.[8] This is the energy stored in the interior of the earth resulting from the gravitational energy generated during the formation of Earth. The primordial heat that had existed before that time is added as a second source. Thirdly, the process of decay of radioactive isotopes on the earth (in particular in the earth's crust) releases heat. Due to the generally low heat conductivity of rocks, this heat resulting from these three sources is to a large extent still stored in the earth.

- Biomass—Releasing solar energy stored in plants and organic matter by burning agricultural waste and other organic matter to generate power, including landfill methane-gas-to-energy conversion. Biomass can be divided into primary and secondary products. The former are produced by direct use of solar energy through photosynthesis. In terms of energy supply, these are farm and forestry products from energy crop cultivation (i.e., fast-growing trees, energy grasses) or plant by-products, residues, and waste from farming and forestry including the corresponding downstream industry and private households (e.g., straw, residual and demolition wood, organic components in household and industrial waste). Secondary products are generated by the decomposition or conversion of organic substances in higher organisms (e.g., the digestion system of animals); these are for example liquid manure and sewage sludge.

- Large hydropower—Of the 80,000 existing dams in the U.S., only 2,400 currently have hydropower generation.[9] These generate power when falling water spins turbines that cause generators to produce electricity. In the U.S., utilities have proposed 70 new projects to upgrade existing generating facilities or to build new capacity. These projects are anticipated to increase hydropower electrical production by 11%, or 11,000 Megawatts over the next decade.[10]

- Small hydropower—Use of flowing water to power electric turbines (small hydro power plants are less than 30 megawatts in size). Of the total solar energy incident on Earth, approximately 21% is used for maintaining the global water cycle of evaporation and precipitation. But only 0.02% out of this amount of energy is finally available as kinetic and potential energy stored in the rivers and lakes of the earth.[11] The water reserves of the earth are available in solid (ice), liquid (water), and gaseous (water vapor) conditions. The water reserve on the earth is continuously cycled by incident solar energy. This global water cycle is mainly fed by evaporation of water from oceans, plants and continental waters. The resulting precipitation which feeds snow fields, glaciers, streams, rivers, lakes, and the ground water is a source for small hydropower. There are thousands of sites in the U.S. with existing dams that are available for mini-hydropower applications. A small hydropower installation at the Kentucky River Lock and Dam No. 2 in Mercer County, Kentucky (the Mother Ann Lee hydroelectric plant), began operation in 2007[12] and will provide power for the equivalent of 2,000 residences.

TECHNOLOGIES FOR WATER CONSERVATION

The goal of efficient water management is to reduce water consumption without compromising the performance of equipment and fixtures. Using water more efficiently is a green strategy. Efficient water management reduces the pressure on limited water resources, the amount of energy and chemicals used for water processing and wastewater treatment, and the quantities of hot water use. These initiatives increase energy savings and the associated environmental benefits, such as GHG and CO_2 emissions reduction. Water conservation also saves money.[13]

Some of the recommended water management techniques include:

- Reducing losses by repairing leaky faucets and pipes.
- Reducing the overall amount of water consumed, such as replacing toilets with low-flush models, etc.
- Finding more sustainable sources of fresh water, such as rainwater harvesting, etc.
- Managing water more responsibly after use, such as using graywater for irrigation, wastewater treatment, etc.
- Enforcing conservation-based water pricing.
- Forming partnerships with local water suppliers and utilities that have the opportunity to provide incentives.

LANDFILL GAS EXTRACTION

A landfill is a solid-waste disposal site. At such locations, waste is generally spread in thin layers, compacted, and covered with a fresh layer of soil each day.[14] The decomposition of the wastes generally generates methane (CH_4) gas. As previously mentioned, CH_4 is a hydrocarbon molecule that is a greenhouse gas. CH_4 has global warming potential estimated to be 23 times higher than CO_2. CH_4 is produced through anaerobic (without oxygen) decomposition of waste in landfills, animal digestion, decomposition of wastes, production and distribution of natural gas and petroleum, coal production, and incomplete fossil fuel combustion. CH_4 is the principal component of natural gas. It is also formed and released to the atmosphere by biological processes occurring in anaerobic environments. Once in the atmosphere, CH_4 absorbs terrestrial infrared radiation that would otherwise escape to space. This property contributes to the warming of the atmosphere, which is why CH_4 is classified as a greenhouse gas. Methane's chemical lifetime in the atmosphere is approximately 12 years.[15] CH_4's relatively short atmospheric lifetime, coupled with its potency as a greenhouse gas, makes it a candidate for mitigating global warming over the near term (e.g., the next 25 years or less).

Projects that transform landfill gas (LFG) to energy significantly reduce bio-methane emissions from municipal solid waste (MSW) landfills—and simultaneously create green energy.[16] MSW landfills are the largest human-generated source of methane emissions in the United States; they released an estimated 38 million metric tons of carbon equivalent (MMTCE) to the atmosphere in 2004. In general, LFG projects could capture roughly 60% to 90% of the bio-methane emitted from landfills, depending on system design and effectiveness. The captured methane is converted to water and less potent CO_2 as the gas is burned to produce electricity. Producing energy from LFG avoids the need to use non-renewable sources such as coal, oil, or natural gas.

Many LFG companies have developed successful operations across jurisdictions that help achieve the goals of sustainability. Waste Management, a company that operates 281 landfills across the U.S., is turning the production of electricity from renewable landfill gases into a major business opportunity. Landfill gases are typically composed of 40% to 50% CO_2 and 50% to 60% CH_4.[17] Landfill gas projects operate about 95% of the time, providing the potential to reduce base-load electrical power and greenhouse gases that would normally be flared.

Waste Management currently operates 103 landfill gas-to-energy (LFGTE) plants. The company plans to create new revenue streams by developing additional sites, with another 60 more to come over the next five years, adding 230 megawatts to its total generating capacity.[18] These projects will be located in Texas, Virginia, New York, Colorado, Massachusetts, Illinois, and Wisconsin. When completed, the company will be a mini-utility. According to Paul Pabor, the company's vice-president of renewable energy, "This initiative is a major step in Waste Management's ongoing efforts to implement sustainable business practices."[19]

Combined with their existing landfill gas generation facilities, the total electrical energy produced when all projects are completed will be 700 megawatts—roughly the equivalent of a fossil-fuel power plant.[20]

ELECTRICAL ENERGY STORAGE

One common problem with electrical generation has to do with having adequate supplies available to meet periods when electrical demand loads are highest. To satisfy the peak load requirements, additional generation must be kept online. Keeping the additional generation capacity online requires energy and, if such energy is carbon-based, this will increase carbon emissions. When demand moderates, generators must still remain on-line, thus producing more power than is needed. Electricity typically needs to be used the instant it is generated. Consider all the GHG emissions that could be reduced if additional generating capacity (new power plants) were not required for peak periods. There is a way to store large amounts of electrical power and save it for peak load conditions.

Sodium and sulfur (NaS) battery banks can store large amounts of electricity. In the past, lead-acid batteries were widely used, even though they required a warehouse-full of interconnected batteries to store electricity. Battery life was only about five years. The lead in these batteries contaminated the environment if disposal was not properly handled. The development of NaS battery banks eliminates the inconveniences and obstacles associated with lead-acid batteries. By bridging electrodes with a porcelain-like material, a room-sized bank of durable NaS batteries will last approximately 15 years. Use of NaS battery banks also reduces the threat of environmental impact. NaS batteries may transform how electricity is stored and delivered to meet peak power demands.[21]

According to Stow Walker of Cambridge Energy Research Associates, by "using NaS batteries, utilities could defer for years, and perhaps avoid, construction of new transmission lines, substations, and power plants."[22] This is possible since the NaS batteries can be charged during periods of low electrical demand, such as nights or weekends when electrical costs are lowest, and then discharge power during peak demand conditions. Green utilities benefit since the battery banks can provide back-up power during power outages and be used in combination with solar and wind power production.[23] In these applications, electricity is stored during peak production periods and released when additional capacity is needed.

Costs for NaS batteries are approximately $2,500 per kilowatt or about 10% more than the cost of building a new coal-fired power plant to produce electricity—but battery systems are much smaller, do not require additional fuel, and do not emit additional greenhouse gasses.[24] NaS battery systems are a proven technology that is already being used by some electric utilities in the U.S. and Japan.

DISTRIBUTED GENERATION

Distributed generation, also known as distributed energy resources (DER), refers to a variety of relatively small and decentralized electrical power generating technologies, such as micro-turbines, fuel cells, and solar photovoltaic (PV). These technologies can be combined with energy management and storage systems, located close to the point at which the electricity is consumed.

DER is more than just a mix of generation technologies; it is the system integration of the generation source, including the energy storage systems and the system that delivers the power-generated interconnection. DER technologies can beneficially replace centralized systems for safety, security, environmental stewardship, cost effectiveness, and efficient operations and maintenance. The system might include components such as generators, the control system, energy storage, and interaction with the electrical power grid. DER technologies include:

- Micro-turbines
- Advanced Industrial Turbines
- Combined Heat and Power or CHP Systems
- Fuel Cells
- Natural Gas Reciprocating Engines
- PV/Solar Systems
- Biomass Systems
- Wind Energy Systems

DER technologies may offer the following benefits:
- Potential source of high-reliability power for sensitive/secure facilities when coupled with an uninterruptible power supply (UPS).

- Greater predictability of energy costs
- Reduction in energy and electrical demand charges.
- Cost-effective source of peak demand power.
- Environmental benefits such as cleaner, quieter operations and reduced GHG emissions since the generators often rely on natural gas or renewable power from solar, wind, biomass, etc.
- Capacity additions can be made more quickly in response to power demands.

In certain circumstances, DER technologies offer widespread benefits to the entire power grid. Examples of such broad benefits include: deferral of new transmission and distribution (T&D) capital investments; reduction of T&D electrical line losses; improved power quality and reliability (voltage support, source of reactive power, and power factor correction); and optimal use of the existing grid's assets such as the potential to increase wheeling capacity. The availability of these benefits depends on the specific site and related energy needs. The particular physical, economic, and regulatory situations, in which the existing centralized electrical grid is operating, impacts the feasibility of such options. Thus, maximizing benefits and their value requires technical, marketing, and policy expertise.[25]

NUCLEAR ENERGY

The urgent need to embrace a new energy future is clear. However, until renewable energy sources, clean coal technologies and major advances in energy efficiencies combine to set the U.S. on the path toward improved energy efficiency and security, there will be growing pressure to advance nuclear power production. This will be particularly important to meet short-term carbon reduction goals.[26] Nuclear energy is carbon-free and it is also available. Nuclear fission and fusion may become a tremendous source of clean energy, without producing GHG and CO_2 emissions. Transformations to useful energy and power are complex.

Many countries already have commitments to develop and use nuclear energy. France, for example, obtains about 78% of its electrical power from nuclear energy. In the rush to embrace nuclear power we must consider three hurdles:

1) The production of nuclear energy is more expensive than new coal or gas-powered plants.
2) Global expansion of nuclear power raises concerns that radioactive nuclear weapons may inadvertently be proliferated.
3) The need to secure and store the resulting radioactive waste for thousands of years.

Nuclear energy may not offer a viable long-term solution until these hurdles are adequately addressed. Despite these drawbacks, 29 new nuclear power plants are being built around the world. The design and development of more than 100 nuclear power plants by various governments is predicted over the next three decades.[27] In particular, India and China are rushing to construct dozens of reactors. As a result, nuclear power cannot be dismissed as part of a response to global climate change.

QUESTIONABLE TECHNOLOGIES

As the response to global climate change and the need to reduce GHG and CO_2 emissions accelerates, proposals for new, innovative and ambitious technologies surface. Many of these will not become realities for decades. Development and commercialization processes require extensive time and funding. However, there are promising clean energy technologies like hydrogen; carbon and CO_2 capture, storage, and sequestration; bio-fuel conversion; and advanced renewable options. At the same time, there are many questionable GHG and CO_2 emissions reduction technologies that are not practical in terms of cost, time, and applications.

One is the proposal is dump CO_2 in oceans near the sea floor. This storage option is quite controversial. At underwater depths of 5,000 meters (roughly 16,400 feet) CO_2 changes its physical state and becomes a liquid. Liquid CO_2 is denser than water and, consequently, would settle on the bottom of the ocean. Forming an undersea lake. Then, a chemical called cathrate would form on top of this lake and store the CO_2 beneath it. According to many scientists and experts, this proposed storage solution is problematic. There is little information available concerning the deep ocean; many fear the impact of carbon storage on this ecosystem. This process would also contravene international ocean-dumping and disposal treaties. Strong public opposition has proven to be a major obstacle.

Nevertheless, there remain ample opportunities for carbon capture technologies. One example is found within the FutureGen project for coal-fired power plants. FutureGen represents a concept for an integrated complex that would both generate power and sequester carbon. Currently some 6 billion tons of CO_2 is emitted into the atmosphere globally (or on the average, one ton per capita per year) from sources such as electricity generating power stations, chemical plants, steel foundries, transportation vehicles—all parts of human activities, which cause GHG and CO_2 emissions to be released into the atmosphere.[28] Developing integrated complexes at locations where concentrated emissions of CO_2 are generated may be one solution.

SUMMARY

This chapter offers an optimistic message by the identifying and discussing recent technological advancements that address GHG emissions. It reiterates the ability of human societies to conduct and analyze research, redirect efforts, redesign infrastructure, and reform lifestyles in order to preserve Earth's ecosystems. There is great potential in the effectiveness of new technologies and practices that can mitigate GHG and CO_2 emissions. Also, there is wide-range in the availability, development and collaboration of technological solutions. Beyond energy conservation and improvements in energy efficiency—the most cost-effective and readily obtainable reduction practices—there are renewable energy resources, GHG storage and manipulation and other technologies that are *available now*. Admittedly, the idea of *availability* is a central theme in this chapter.

Renewable energy resources that emit little or no air pollution are naturally replenished on Earth and are *available*. Moreover, there have been impressive gains in the efficiency with which energy is utilized for heating, lighting, refrigeration, and power for industry and transportation. Technological innovation is rapidly developing. The 20th Century has seen the successful commercialization of products and services such as compact fluorescent light bulbs (CFLs) and the substitution of natural gas for coal to the generate electrical power. Other green strategies include water management and conservation, landfill gas extraction, the development of nuclear energy, and DER technologies. Altogether, the technologies described in this chapter demonstrate a beneficial, strategic diversification of energy sources and technological improvements. Further expansion of alternative, environment-friendly technologies will provide the world with an increase in the flexibility, adaptability, and viability of carbon mitigation efforts.

Endnotes

1. President John F. Kennedy (1962, September 12). Address at Rice University on the nation's space effort. http://www.jfklibrary.org, accessed 28 May 2008.
2. FutureGen Alliance (2008, July 11). *FutureGen Alliance hails Senate Appropriations Committee for protecting FutureGen at Mattoon Funding.* http://www.futuregenalliance.org/news/releases/pr_07-11-08.pdf, accessed 29 December 2008.
3. Renewable energy investment skyrockets (2008, September). *Cities Go Green.* p. 6.
4. An Apollo program for climate change (2007, June 22). *The Washington Post.* p. 19.
5. Davidson, P. (2008, September 8). 4 creative solutions to energy problems. *USA Today.* p.4B.
6. Kaltschnitt, M., Streicher, W. and Wiese, A. (2007). *Renewable energy: technology, economics and environment.* Springer-Verlag.
7. *Ibid.*
8. *Ibid.*
9. Davidson, P. (2008, October 28). Water power gets new spark. *USA Today.* p. 33.
10 *Ibid.*
11. *Ibid.*
12. Bruggers, J. (2008, October 28). Retrofit hydro plant offers clean power. *The Courier-Journal.* p. E1.
13. U.S. Department of Energy (2001). *Greening federal facilities: An energy, environmental and economic resource drive for federal facility managers and designers* (2nd ed.).
14. See Glossary of terms. http://www.nea.gov.vn/html/VEM_2004/69-77_Eng.pdf, accessed 23 January 2009.
15. U.S. EPA (2008, October 19). *Greenhouse Gas Properties.* http://www.epa.gov/methane/scientific.html, accessed 23 January 2009.
16. Renewable Energy Technologies Company
17. Renewable Energy Access (2007, June 27). *700 MW from electricity to come from landfill gas.* http://www.renewableenergyaccess.com/rea/news/story?id=49123, accessed 7 July 2007.
18. *Ibid.*
19. *Ibid.*
20. *Ibid.*
21. Davidson, P. (2007, July 5). New battery packs power punch. *USA Today.* p. 3B.
22. *Ibid.*
23. *Ibid.*
24. *Ibid.*
25. U.S. Department of Energy (2006). DER/CHP. *Federal energy management program.* http://www1.eere.energy.gov/femp/, accessed 28 May 2008.
26. Kerry, J. and Kerry, T. (2007). *This moment on Earth: Today's new environmentalists and their vision for the future.* New York: Public Affairs.
27. An Apollo program for climate change (2007, June 22). *The Washington Post.* p. 19.
28. Intergovernmental Panel on Climate Change (2007, November). *Fourth assessment report.* http://www.ipcc.ch/ipccreports/ar4-syr.htm, accessed 28 May 2008.

Chapter 8

Carbon Sequestration Technologies

By Stephen Roosa & Danielle Miller

"Carbon sequestration, also known as carbon capture and storage (CCS), is a technological approach to solving global warming by preventing the CO_2 from entering the atmosphere when coal is burned. The process involves capturing CO_2 at coal-fired power plants, transporting the gas, and storing it instead of letting it enter the atmosphere. However, it is extremely expensive—not nearly cost-competitive with other solutions to the climate crisis. There are also major concerns over CO_2 leakage and the fact that the sequestration process itself requires large amounts of energy. Additionally, the technologies needed for the process are not likely to be ready in time to apply CCS to new power plants about to come online, and retrofitting old plants with CCS is very expensive."

Co-op America[1]

Carbon sequestration is "the placement of CO_2 into a repository in such a way that it will remain permanently sequestered."[2] Carbon sequestration is not a new phenomenon. In fact, it is a natural process that occurs throughout various ecosystems. Forests and oceans are the most successful examples, storing large quantities of carbon. Remarkably, the quantity of carbon within the Earth's natural cycle of land, ocean and air exchanges is ten times the rate of annual anthropogenic emissions.[3] However, these natural processes do not have the ability to immediately process the growing amount of anthropogenic carbon emissions and, consequently, the carbon-content in the atmosphere increases.[4] To effectively reduce the magnitude of atmospheric carbon, a variety of complementary measures and manipulations are required. Terrestrial sequestration involves the use of vegetation and soils as carbon sinks. Thus, the enhancements of natural processes (e.g., afforestation or increasing photosynthesis in the oceans) as well as the development of technological sequestration designs (e.g., subsurface storage for carbon) offer enormous potential for carbon reduction.[5] These are successful when they can either prevent the release of CO_2 into the atmosphere or effectively store it elsewhere.

Geological sequestration involves the "permanent storage of CO_2 in geological formations below the earth's surface."[6] One example is a project off the coast of Norway that involves removing CO_2 from natural gas and injecting it into a saline reservoir under the North Sea.[7]

ENHANCEMENT OF
NATURAL PROCESSES AND GEOENGINEERING

There are a number of ways to reduce atmospheric carbon through the direct manipulation of the natural environment. Enhancing the capacity of the CO_2 cycle involves both biological and ecological processes that capture and remove carbon from the atmosphere. Vegetation and storage in biomass and soils provide excellent examples. This can be achieved by increasing the rates of photosynthesis using vascular plant life, finding ways to retain carbon in soils, preventing adverse land-use changes, and increasing the capacity of deserts and degraded lands to sequester carbon.[8] Reforestation and afforestation also contribute to mitigation efforts. Indeed, forested lands can process a substantially higher amount of CO_2 in comparison to fallow lands. As a result, protecting forested regions from damage due to development can be a potent carbon reduction strategy. In 2005, Bolivia's Noel Kempff Climate Action Project, which protects 3.8 million acres of tropical forest, became "the first conservation-based initiative in the world to be fully certified for reducing greenhouse gas emissions."[9] From 1997 to 2005 the project prevented the addition of an estimated 1,034,137 metric tons of CO_2 by disabling the rights for timber harvesting, logging, and other types of deforestation.[10]

Another possible technique that would enhance carbon sequestration within the earth's oceans is the fertilization of phytoplankton. Research shows that the addition of micro and macronutrients to the ocean surface instigates a dramatic growth in the population of small plants—causing phytoplankton to multiply their numbers two to three times in a single day.[11] Subsequently, phytoplankton can extract CO_2 from the atmosphere and transport carbon as dead organisms sink to deeper ocean levels.[12] Though most scientists agree that the fertilization of the ocean surface would decrease carbon emissions in the atmosphere, some hesitate to support this carbon sequestration practice. Such "geo-engineering" solutions could have unanticipated impacts on the more delicate ocean ecosystems. Concerns arise over the threat to disrupting food-web dynamics, carbon cycle dynamics, silicon dynamics, and calcium carbonate dynamics.[13] Moreover, further research and validation is required before any large-scale investments in ocean fertilization are likely to occur. Validation is possible through pilot-testing bio-chemical methodologies.

Successful geo-engineering projects intended to enhance carbon sequestration processes depend on technological advancements and increased awareness of the environmental risks associated with human activities. Examples include algae farms and enhanced oil recovery. Interestingly, technology originally developed by the human genome project is now being applied to study microorganisms that use carbon to produce either hydrogen or methane. New research targets their genetic structures, trying to identify the components that enable and maximize the capture and use of carbon in the organism.[14] With the appropriate knowledge and research results, scientists may be able to engineer biological entities to ingest fossil fuels or carbonaceous sources.[15] However, one fear is that such approaches may yield unintended consequences or immeasurable results.

CARBON SEQUESTRATION TECHNOLOGIES AND PROCEDURES

In addition to natural processes, technological methods exist and are being developed. Most technology-based carbon sequestration methods are centered on the carbon capture and storage (CCS) process. Indeed, the difference in most sequestration projects is the nature of geographical storage structures (e.g., saline formations, depleted oil and gas reservoirs, deep and unmineable coal seams) not the process for retrieving and purifying the carbon.[16] CCS includes the following steps:

1) Carbon must be captured from an emission source
2) The stream is cleaned of extraneous molecules and compressed
3) The compressed carbon is transported and stored

From this list, several important factors of technological carbon sequestration can be discerned. First and foremost, CCS typically relies on expensive, stationary infrastructure to clean, compress and transport the carbon. The process also requires large, carbon-rich emissions in order to be cost-effective. Accordingly, most CCS options are confined to large, point source emitters with high carbon-content streams: fossil-fueled power plants, chemical plants, refineries, iron and steel foundries, cement plants, and natural gas processing sites.[17] Stationary sources of carbon emissions are well documented by international organizations and research groups. Worldwide, more than 14,600 point sources have been identified. These tend to be grouped into three large clusters and located in the following industrialized regions: the mid and eastern states in the U.S., the central region of Europe, and in South East Asia (i.e., China and Japan).[18]

The economic and political feasibility of a carbon sequestration project is heavily influenced by the source's proximity to a sound, tested geological structure. The predicted length of storage time is also quite important. Some of the "geological trap types" under consideration were reservoirs that successfully confined hydrocarbons or liquids for millions of years.[19] However, there is great uncertainty about the possible environmental and health effects should a storage facility leak carbon into underground water systems or into the air.[20] Indeed, the capacity of the geological structure and its distance to a source of carbon emissions often dictates whether or not the project will be undertaken. Co-location of facilities is often a key to success. Co-location is a planning solution whereby facilities that normally generate carbon emissions are located in close proximity to repositories that are capable of separately using and storing the carbon.

The most promising and lucrative technologies not only prevent the release of carbon into the atmosphere, but also integrate the process with industry in economically attractive ways. Benefits accrue when a functional use of the captured carbon is found.

ENHANCED OIL RECOVERY

Carbon can be used in enhanced oil recovery (EOL) processes. Recovering oil from underground reservoirs using conventional techniques involves drilling and using pressure and pumping systems to bring oil to the surface. This method of recovery is not entirely effective and leaves residual oil that is not economically recoverable with conventional technologies. To enhance recovery, water can be pumped into the oil-bearing rock, increasing the pressure and allowing more oil to be recovered. The increased pressure allows the oil-bearing liquids to be pumped out. Then the oil is separated and recovered. This process typically consumes large quantities of water, which can seep into aquifers and cause contamination.

There are large reserves of oil that remain from previously developed oil fields, often with abandoned wells that have been capped. This creates the opportunity to use high pressure CO_2, rather than water, for EOL. There are a number of benefits when CO_2 is used. Since water use is substantially reduced, the potential of contaminating groundwater reserves is virtually eliminated. When CO_2 is injected into the oil-bearing rock, the process sequesters the carbon in locations where it is unlikely to be released into the atmosphere. CO_2 scrubs molecules of both oil and natural gas from rock formations much more effectively than water. Thus, greater quantities of oil can be recovered.

This process is a proven, albeit expensive, technology to deploy. It is currently being used at the Great Plains Synfuels Plant (owned by the Dakota Gasification Company) in North Dakota. This plant produces synthetic natural gas from lignite coal and uses CO_2 as a commercially viable by-product for enhanced oil recovery. The CO_2 generated there (95 million cubic feet daily) is piped 330 km (205 miles) to the EnCana Corporation's Weyburn Oil Fields in Saskatchewan, Canada.[21] There it is injected into depleting oil formations to increase oil production, a form of geological sequestration, at a rate in excess of one million metric tons annually.[22] Similar methods are also being developed elsewhere. In fact, EU estimates indicate that available underground storage capacity is equivalent to seven hundred years of CO_2 emissions from the EU's total electrical production.[23]

ZERO EMISSION COAL TECHNOLOGIES

The use of coal accounts for approximately 40% of all CO_2 emissions worldwide.[24] In the U.S., a country with 27% of the world's known coal reserves, coal-fired power plants produce about half of the annual electrical energy used. Moreover, coal is expected to fuel 55% of the country's electrical production by 2025. With about 20 more of these plants under construction, their generation of carbon emissions will become increasing important. The cost of developing coal-fired generating facilities is increasing. Yet the availability of coal as a resource is too great to ignore. In the U.S., Peabody Coal has reserves that are 50% greater in energy content than all of Exxon's oil reserves. The company and its partners are constructing a $3 billion 1,600 MW electrical power plant in Illinois that will be the largest constructed in the U.S. in over 25 years.[25] This is roughly $1,875,000 per MW of electricity produced. With coal usage already accounting for 84% of the total carbon emissions that are generated in the U.S. and climate change legislation likely in the near future, U.S. utilities have recently cancelled or delayed construction of an additional 60 coal-fired power plants.[26]

However, it is possible to burn coal with only negligible carbon emissions impact. Coal might be a bridge to the future—a potential energy supply—if ways can be found to encourage its use and reconcile its harmful environmental impacts.

Coal-Fired Electrical Generating Facility, Evansville, Indiana

One seemingly utopian concept is the idea of the zero emissions coal (ZEC) power plant, which combines an electrical generation system with a sequestration system. A ZEC power plant might use coal slurry and could reduce carbon emissions by combining a set of processes:[27]

1) Gasification process whereby carbon (e.g., from coal slurry) is gasified to produce methane and steam.

2) Carbonation process in which the methane and heat produced by gasification are fed into a carbonation reactor in the presence of water, steam and calcium oxide (lime);

3) Calcination process that uses waste heat from electrical generation in a fuel cell system to regenerate lime for carbonation producing pure CO_2 that can be sequestered;

4) Electrical generation using a solid oxide fuel cell which results in a higher than normal energy conversion efficiency (roughly 70%); and

5) Carbon sequestration method that combines CO_2 with mineral carbonates, such as serpentine or fosterite, than can be stored in surface or underground mine-openings.

Geological sequestration of carbon dioxide will require injecting carbon deep underground without concerns that it might taint groundwater or eventually escape to the surface and ultimately into the atmosphere. Ben Grumbles from the U.S. Environmental Protection Agency describes this technology as "promising yet unproven."[28] Sequestration will require careful study of storage sites, assurance that storage is permanent, and funding for well plugging, site care and emergency response.[29] A regulated rule structure will need to be developed and adopted by the federal government which is palatable to states with coal burning electrical utilities.

In theory, ZEC offers the benefit of using coal in a manner that prevents release of carbon into the atmosphere; it reduces the harmful environmental impacts normally associated with using coal for combustion processes. The benefits of ZEC technologies support the rationale that "clean coal" electrical production is possible. However, the implementation of such sophisticated technologies will likely be expensive.

CARBON SEQUESTRATION IN OCEANS

The world's oceans have the potential to store large quantities of anthropogenic carbon emissions. Two strategies for enhancing carbon sequestration in the oceans include: 1) enhancing the net oceanic uptake of CO_2 molecules from the atmosphere by fertilization of phytoplankton with micro or macronutrients, and 2) injecting a relatively pure CO_2 stream to ocean depths greater than 1,000 meters.[30] At greater depths it may be possible to store it on the bottom of the ocean as a liquid or icy hydrate deposit. The effectiveness and potential environmental consequences of ocean sequestration by either strategy are unknown and unproven.[31]

A third, related method is storing CO_2 in geological formations below the ocean floor. The EU initiated one of the first industrial-scale projects, The Saline Aquifer CO_2 Storage Project, using this kind of storage location in 1996. The Saline Aquifer CO_2 Storage Project is located in the North Sea's Sleipner Field, a saltwater-bearing sandstone formation that is one kilometer below the sea floor. Thus far, the project has provided storage for over 1 million tones of CO_2.[32]

PROCESSING CO_2 USING ALGAE

Algae are unicellular plants that thrive on carbon dioxide, ingesting it while releasing oxygen and producing economically important byproducts. This photosynthetic process occurs in the presence of sunlight. Algae have thrived on Earth for billions of years. While water is required, algae do not require potable water to proliferate. Algae is laden with combustible vegetable oils that can be harvested for use in the production of bio-diesel fuels, paper or plastic products, and starches that can be used to produce ethanol. Algae are also a protein source that can be used to manufacture fish and cattle feed.[33]

Dr. Isaac Berzin, while at Massachusetts Institute of Technology (MIT), proved the viability of a hypothetical 20-MW power plant. The company Dr. Berzin founded in Cambridge, Massachusetts, GreenFuel Technologies, garnered $11 million in venture capital to construct an algae bioreactor system and to successfully conduct a field trial in Arlington, Arizona, at an operating power plant. Bolted to the power plant's exhaust stacks are rows of tubes that contain a rich green algae soup that consumes much of the power plant's CO_2 emissions. Altogether, it reduces CO_2 emissions by 40% and nitrous oxide emissions by 86%.[34] Specially designed pipes capture the CO_2 from the stacks and serve as conduits to direct the gas to containers that hold the live algae in the presence of sunlight. CO_2 is consumed in the process and oxygen is generated. The algae can be harvested daily. Today, "algae growing exponentially and absorbing vast quantities of carbon dioxide from smokestacks at Arizona Public Service Company's (APS) Redhawk electric plant is being billed as an answer for greenhouse gases and a source of biodiesel and ethanol."[35] Over 100 types of algae are suitable; however, the use of algae with a high oil density (50% by weight) is important to the success of the process.

Though a large amount of land is needed for an "algae farm," it is much less than the land area required for growing corn to produce ethanol. It requires approximately 0.8 hectares (two acres) of algae to process the carbon generated by one megawatt of electricity produced by a typical coal-fired power plant. Berzin estimates with his design a single 1,000-MW power plant with an 810-hectare (2,000 acre) farm of algae filled tubes could produce 181 million liters (40 million gallons) of biodiesel and 227 million liters (50 million gallons) of ethanol annually.[36]

AGRICULTURAL METHANE CAPTURE

Agricultural wastes (from pig stock, cattle production and dairy farming) produce large quantities of manure. When this manure is stored in fields, open tanks or lagoons, CH_4 gas is a natural by-product that is released directly into the atmosphere as the waste materials decompose. CH_4 is a very harmful greenhouse gas that breaks down in the atmosphere and generates CO_2.

There ways to prevent or reduce the addition of GHG to the atmosphere from agricultural practices. For example, a bio-fueled plant in Sandbeiendorf, Germany, processes these wastes—converting them into heat and electricity—while reducing CH_4 emissions by 90%.[37] Not only are CO_2 emissions reduced but the facility also produces much of its own heat and power. A similar project in Andhra Pradesh, India, uses agricultural biomass and wastes to generate electricity for local industries, providing power that would otherwise need to be purchased from the state grid.[38]

CH_4 is also produced when agricultural wastes are placed into landfills. Composting organic wastes using aerobic methods substantially reduces or eliminates methane gas production, producing CO_2 instead. In this case CO_2 is a preferable alternative and lowers the global warming potential of the emissions that are generated. An organic waste composting project in Victoria, Australia, does just this, reducing carbon emissions by 15,000 metric tons annually.[39]

ALTERNATIVES TO SEQUESTRATION

Alternative technologies offer promise that sustainability and carbon reduction goals can be accomplished. There are industrial uses for carbon that provide alternatives to sequestration. Carbon dioxide is used by the beverage industry in canning and bottling processes for products. Carbon fibers can be manufactured for use in structural applications, binder and cokes for aluminum and activated carbons have adsorption characteristics that hold great promise.[40] Carbon can also be used as a product in the manufacture of high strength, highly conductive composite materials in electromagnetic shielding applications and electrochemical energy storage devices.[41] Recycled biocatalysts (such as enzymes and bacteria) can be used to break down carbon dioxide and water.[42] The carbon can be combined with hydrogen to create methanol that can be used as a substitute for liquid fuels such as oil. Carbon Sciences, a company in Santa Barbara, California, believes that while energy is needed for this low-temperature, low-pressure process, it may become economically viable.[43]

According to the researchers at the Georgia Institute of Technology, processes capable of capturing carbon dioxide emissions from large-scale sources, such as power plants, have recently gained some impressive scientific ground. However, nearly two-thirds of global carbon emissions are created by much smaller polluters—automobiles, transportation vehicles, and distributed industrial power generation applications (e.g., diesel power generators). The Georgia Tech team hopes to improve transportation system sustainability by using a liquid fuel and separating, trapping and storing the carbon emissions generated by the vehicles. The carbon would then be shuttled back to a processing plant where it would be transformed into liquid fuel.

Perhaps the best way to avoid the costs and financial risks associated with sequestering carbon is to reduce the use of fossil fuels altogether. This is most readily accomplished by energy efficiency, energy conservation or fuel substitution (using alternative energy solutions). Electrical power plants contribute over 25% of the world's CO_2 emissions. In 2006, Airtricty, a Dublin-based developer of wind power, announced plans for a 10,000 megawatt European supergrid that will use 2,000 offshore wind turbine generators (Kaihla 2007:69). The electrical production would be equivalent to the output of several conventional power generating plants. When compared to energy from traditional sources, CO_2 emissions will be reduced by the equivalent of 60 million tons per year.

There are multiple examples of how providing incremental improvements in technologies that are widely used can ultimately reduce the need for carbon fuels. Technologies such as CFL lighting and high efficiency electric motors are often cited as examples. While energy-efficient electric motors may be slightly more expensive to purchase and install, the savings can be substantial. In the industrial sector, electric motors use as much as 60% of total electrical energy. Energy efficient electric motors can provide a 2% to 6% efficiency improvement over standard motors. Such replacements can reduce electrical loads, life cycle costs and ultimately carbon emissions. While a more efficient motor may cost slightly more than a standard motor, the savings can be substantial. An estimated "97% of the life cycle costs of a standard motor goes to energy costs and only 3% to procurement and installation."[44] Yet in the upside-down world of purchasing practices, infrastructure components are often acquired based on lowest initial costs.

A number of companies are developing technologies that use plasma-arc processes to eliminate wastes and create interesting by-products. Prototype systems are already in operation. These plants are a means of converting post-consumer municipal solid wastes into usable products. Plasma-arc gasification uses a high temperature (1,650°C or 3,000°F), lightning-like process that breaks down shredded solid waste into molecular building blocks which provide marketable by-products: 1) a combustible synthetic gas (syngas); 2) metal ingots; and 3) a glass-like solid that can be used for floor tiles or gravel (Durst 2007). Materials that can be processed using plasma furnaces include municipal and household wastes and hazardous wastes such as oil sludge and those generated by hospitals. Plasma-arc incineration plants have been constructed in Yoshii, Japan (2000), Utashinai City, Japan

(2002), Mihama, Japan (2002), and Ottawa, Canada (2007).[45] However, dioxin can be emitted and chlorine is often found in the wastes.[46] Plants using this technology in Australia and Germany were shut down because they were unable to meet local emission standards.[47] Carnival Cruise Lines uses a similar system developed by PyroGenesis to process five tons of waste daily on one of its ships, reducing the waste to a few kilograms of sand (Durst 2007).

St. Lucia County in Florida is on the verge of having the first plasma arc incineration facility in the U.S. Landfill wastes will be gasified to yield synthetic gas, slag for road construction and steam for a nearby factory.[48] In addition, the $425 million project will provide 120 megawatts of electrical power using turbines, one-third of which will be used to operate the facility.[49] Tropicana Products, Inc., will purchase the 80,000 pounds of steam generated daily for its juice plant's turbines.[50] Faced with an accumulation of 4.3 million tons of trash collected over 30 years, increasing waste disposal costs, and a shortage of landfill space, the county plans to resolve all of these problems by eliminating the need for the landfill. When completed, this plant will vaporize 3,000 tons of garbage per day, enough to consume the wastes in the county's landfill within 18 years.[51] While the process emits CO_2, it is much less than the amount produced from traditional carbon-based fuel sources. According to Louis Circeo, director of Georgia Tech's plasma research department, this technology "could not only solve the garbage and landfill problems in the U.S. and elsewhere, but it could significantly alleviate the current energy crises."[52]

Technologies are widely available today to reduce energy consumption and cut our dependency on fossil fuels. Many offer economic benefits. In most cases, trade-offs are necessary. U.S. anthropogenic emissions of greenhouse gases have increased 16.9% from 6,113 metric tons in 1990 to 7,147 metric tons in 2005 and are not subsiding.[53]

Renewable energy will be an important part of our portfolio of sustainable energy solutions. Sustainable energy development has its own "upside-down" economics. While solar and wind power often have higher initial costs, they also have negligible carbon impact, and there is no need to purchase fuel. Conventional fuel solutions have lower initial costs, burdensome carbon emissions, and must use ever-more costly fuels. As the right to emit carbon into the atmosphere becomes more costly and non-carbon energy sources become less expensive, alternative energy solutions will become more viable.

SUMMARY

What can be done regarding unaddressed environmental problems such as CO_2 gas emissions? Carbon capture and sequestering technologies are becoming available and offer hope. Perhaps the most efficient "machine" to capture and store atmospheric carbon are the natural processes that over millions of years successfully turned carbon into the oil, coal and other minerals that we use today. In a manner, technologies that sequester carbon attempt to mimic those natural processes.

CO_2 currently being released into the atmosphere can be stored in oceans, in caves, abandoned oil and natural gas fields, and elsewhere. Not all of these sequestration techniques are environmentally friendly. When stored in water, carbonic acid can be formed, affecting marine life. Though costly, the use of liquid sodium hydroxide (lye) holds promise as it has the ability to absorb and capture CO_2 from airstreams. It is most feasible for locations where emissions are concentrated and are being exhausted, such as the stacks of coal-fired power plants.

Terrestrial sequestration involves the use of vegetation and soils as carbon sinks. Planting trees and other forms of vegetation is perhaps one of the most cost-effective solutions—they are natural systems that not only store carbon but also generate oxygen. Geological sequestration involves storing CO_2 in geological formations below ground. A project off the coast of Norway removes CO_2 from natural gas and injects it into a saline reservoir under the North Sea. The Great Plains Synfuels plant in North Dakota produces synthetic natural gas from lignite coal and uses CO_2 as a commercially viable by-product for enhanced oil recovery.

Renewable energy will be an important part of our portfolio of sustainable energy solutions. While solar and wind power often have higher initial costs, they also have negligible carbon impact, and there is no need to purchase fuel. Conventional fuel solutions have lower initial costs, burdensome carbon emissions, and must use ever-more costly fuels. As the right to emit carbon into the atmosphere becomes more costly and non-carbon energy sources become less expensive, alternative energy solutions will become more viable.

Endnotes

1. Co-Op America (2005). *Carbon sequestration.* http://www.coopamerica.org/programs/climate/dirtyenergy/coal/ccs.cfm, accessed 17 April 2008.
2. National Energy Technology Laboratory. *What is carbon sequestration?* http://www.netl.doe.gov/technologies/carbon_seq/FAQs/carbon-seq.html, accessed 17 April

2008

3. Bryant, S. (2007, September). Geologic CO_2 storage—can the oil and gas industry help save the planet? *Distinguished Author Series.*
4. Hultman, N. (2007). Carbon sequestration. *Encyclopedia of energy engineering and technology.* Taylor and Francis, Inc.
5. *Ibid.*
6. Eliason, D. and Perry, M. (2004, October). *CO_2 recovery and sequestration at Dakota gasification company.* http://www.gasification.org/Docs/Conferences/2004/11ELIA_Paper.pdf, accessed 26 December 2008. p. 9.
7. *Ibid.*
8. U.S. Department of Energy. *Carbon sequestration focus areas.* http://cdiac2.esd.ornl.gov/scienceman.html#enchancing, accessed 20 January 2008.
9. The Nature Conservancy. (2005, December). *First conservation initiative certified for reducing greenhouse gas emissions.* http://www.nature.org./initiatives/climatechange/press/press2192.html, accessed 12 June 2008.
10. *Ibid.*
11. Markels, M. and Barber, R. (2001, May 14). *Sequestration of CO_2 by Ocean Fertilization.* Presentation for NETL Conference on Carbon Sequestration.
12. Lawrence Berkeley Laboratory. *Ocean fertilization.* http://www-esd.lbl.gov/CLIMATE/OCEAN/fertilization.html, accessed 12 June 2008.
13. *Ibid.*
14. U.S. Department of Energy. *Carbon sequestration focus areas.* http://cdiac2.ornl.gov/scienceman.html, accessed 10 June 2008.
15. *Ibid.*
16. U.S. Department of Energy (2007, April). *Geologic sequestration research.* http://www.fossil.energy.gov/programs/sequestration/geologic/, accessed 10 June 2008.
17. Ambrose, W., et al. Source-sink matching and potential for carbon capture and storage in the gulf coast. *The Gulf Coast Carbon Center.* http://www.beg.utexas.edu/environqlty/co2seq/pubs_presentations/UIC_Ambrose.pdf, accessed 26 December 2006.
18. Gale, J. (2002). *Overview of CO_2 emissions sources, potential, transport and geographical distribution of storage possibilities.* IPCC Workshop on Carbon Dioxide Capture and Storage.
19. *Ibid.*
20. *Ibid.*
21. Eliason, D. and Perry, M. (2004, October) *CO_2 recovery and sequestration at Dakota gasification company.* http://www.gasification.org/Docs/Conferences/2004/11ELIA_Paper.pdf, accessed 26 December 2008. p. 1.
22. *Ibid.*, p. 11.
23. Mega, V. (2005). *Sustainable development, energy and the city: A civilization of visions and actions.* New York: Springer.
24. Davidson, P. (2008, August 19) Coal king Peabody cleans up. *USA Today.* p. 1B.
25. *Ibid.* p. 2B.
26. Energy Information Administration (2004, February). *Annual energy outlook 2004 with projections to 2025.* DOE/EIA-0554.
27. Yeboah, F., Yegulalp, T. and Singh, H. (2007). Future zero emissions carbon technology—valuation and policy issues. *Energy Engineering.* 104 (6). pgs. 10-12.
28. Bruggers, J. (2008, July 16). EPA proposes first carbon storage rules. *The Courier-Journal.* p. D.1.
29. *Ibid.*, p. D.2.
30. Yeboah, F., Yegulalp, T. and Singh, H. (2007). Future zero emissions carbon tech-

nology—valuation and policy issues. *Energy Engineering.* 104 (6). pgs. 10-12.
31. *Ibid.*
32. Mega, V. (2005). *Sustainable development, energy and the city: A civilization of visions and actions.* New York: Springer.
33. Shafffer, M. (2006, October 14). Algae could be fuel of the future. *The Arizona Republic.* http://www.azcentral.com/arizonarepublic/business/articles/1014biz-algae1014.html, accessed 13 January 2008.
34. Clayton, M. (2006, January 11). Algae – like a breath mint for smokestacks. *The Christian Science Monitor.* http://www.csmonitor.com/2006/0111/p01s03-sten.htm, accessed 12 January 2008.
35. Shafffer, M. (2006, October 14). Algae could be fuel of the future. *The Arizona Republic.* http://www.azcentral.com/arizonarepublic/business/articles/1014biz-algae1014.html, accessed 13 January 2008.
36. Clayton, M. (2006, January 11). Algae – like a breath mint for smokestacks. *The Christian Science Monitor.* http://www.csmonitor.com/2006/0111/p01s03-sten.htm, accessed 12 January 2008.
37. HSBC (2007). *Reducing emissions.* http://www.hsbccommittochange.com/environment/hsbc-case-studies/carbon-dioxide/reducing-emissions/index.aspx, accessed 28 May 2008.
38. *Ibid.*
39. Ecofys UK (2007, February 19). *Carbon offsetting, a report for the Arts Council England.* http://www.arts.org.uk/documents/projects/phpVg6Scb.pdf, accessed 28 April 2008.
40. University of Kentucky, Center for Applied Energy Research. *Carbon materials.* Lexington, KY.
41. *Ibid.*
42. Davidson, P. (2009, February 25). Greenhouse gas villain released. *USA Today.* p. 3B.
43. *Ibid.*
44. Siemens AG. (2007). *Combat climate change – less is more.* http://wap.siemens-mobile.com/en/journal/story2.html, accessed 3 July 2007.
45. Wikipedia. *Plasma arc waste disposal.* http://en.wikipedia.org/wiki/Plasma_arc_gasification, accessed 7 July 2007.
46. Muller, P. *Syngas in Johnson County.* http://mypc.press-citizen.com/story.php?id_stories=304, accessed 7 July 2007.
47. *Ibid.*
48. Zaks, D. (2006, September 12). *Geoplasma – plasma arc incineration.* http://www.worldchanging.com/archives/004926.html, accessed 7 July 2007.
49. *Ibid.*
50. Florida county plans to vaporize landfill trash. (2006, September 9). *USA Today.* http://www.usatoday.com/news/nation/2006-09-09-fla-county-trash_x.htm, accessed 7 July 2007.
51. *Ibid.*
52. *Ibid.*
53. Energy Information Administration (2005). *Emissions of greenhouse gases in the United States,* executive summary. www.eia.doe.gov/oiaf/1605/ggrpt/pdf/executive_summary.pdf, accessed 30 July 2006.

Chapter 9

Corporate Carbon Reduction Programs

"You have to manage your risks before you can manage your carbon... you've got to build it into your strategy and build it into your culture."

Stangis, D. (2007, September 14)
Director of Corporate Responsibility for Intel Corporation.[1]

Corporations have an opportunity to take a leadership role in the reduction of carbon emissions. They can modify their internal policies and goals, implement corporate programs, and place demands on their suppliers. Like many other corporations, Dell Computer Corporation has set its goal to be a green technology company. The company intends to reduce its carbon intensity globally by 15% by 2012.[2] Most importantly, Dell is using this initiative as an opportunity to change the types of products they bring to market. To be successful, product changes must satisfy the needs of the marketplace and the demands of the customers. In order to justify an investment in newer, less carbon intensive and less carbon consuming technologies, corporations such as Dell need to believe that their investments will procure a satisfactory financial return.

Today, there is increasing recognition in financial markets that green companies can be much better investments than many smokestack companies. Roughly 25% of all Fortune 500 companies have a board committee that oversees environmental issues, compared to fewer than 10% only five years ago.[3] Sustainability "has become a much more important part of every board's activities," according to Lester Hudson, chairman of American Electric Power Company's governance committee.[4] There is international competition for venture capital in "clean tech" industries that produce technologies capable of developing clean energy resources. With the economic development stakes increasing, many localities have incentives in place to develop and attract these industries. In 2006, such U.S. companies were funded with roughly $2.8 billion of new venture capital.

The competition to attract "clean-tech" industries is fierce. Three localities collectively attract almost 40% of the clean-tech industries in the U.S., namely, Silicon Valley ($683 million), Boston ($250 million) and Austin ($210 million).[5] Austin, in partnership with the University of Texas, established a clean energy incubator and located start-up capital for 18 green companies.[6] In fact, one company in Austin is developing a 125-acre park to recruit clean-energy companies—the first to commit was a company that manufactures fuel pellets from recycled paper.[7]

Many corporations lack the internal skills to understand what actions need to be taken and how to measure the impacts. This assessment, sorting and prioritizing process may involve fundamental changes in organizational behavior as well as infrastructure modifications. Once decisions and investments are made, validating the investments in green technologies can be challenging for corporations.

THE AUTOMOBILE INDUSTRY

One way to reduce CO_2 emissions is to use less fuel for transportation purposes. Changes in automobile design are driven by state regulations such as those in California (adopted verbatim by at least 11 other states) that require manufacturers to reduce CO_2 emissions and increase fuel economy standards. In response to anticipated changes in the marketplace, the automobile industry continues to develop prototype vehicles that use new, less carbon intensive technologies. Some recent examples include:

- General Motors—Project Driveway is a planned two-year test of a fleet of 100 Chevrolet Equinox fuel-cell vehicles scheduled to be marketed beginning in 2008.[8] Since there are few hydrogen filling stations in the U.S., the company plans to provide temporary filling sites. General Motors will also offer the vehicles at no charge to users in Los Angeles, New York and Washington, D.C., for periods of up to three months to test the market for the vehicles.[9] After having invested millions in developing a new lithium-ion battery system that allows a vehicle to go 64 km (40 miles) on electricity before a gasoline motor begins the recharging process, the company is reinventing the electric vehicle.[10] The Chevy Volt, concept car developed by a team led by Bob Lutz was unveiled in 2007.[11] Unlike many previous electric vehicles, its innovative rechargeable electric drive system and range-extending power source can be configured to run on electricity, gasoline, E85 or biodiesel—all made possible by an innovative propulsion system.[12] If driven 95 kilometers (59 miles) daily, the vehicle is estimated to achieve 24.6 km/l (150 mpg).

- Honda—Using environmentally responsible technology at Honda is what they call "Environmentology." The Union of Concerned Scientists named the company the "Greenest Automaker" in 1989, 2001 and 2003. Having had success with its hybrid vehicles, the company developed a new, hydrogen-fueled vehicle call the FCX and is working on a more fuel-efficient compact hybrid that is scheduled to be launched in the U.S. by 2009.[13] The company has also developed a flex-fuel (using ethanol) and a diesel car that can meet California's 2009 air-quality standards.[14]

- Hyundai—The South Korean vehicle manufacturer is developing the prototype hydrogen sedan "i-Blue" and has converted over 300 Tucson SUVs to operate on fuel cells.

- Mercedes-Benz—The German automaker has developed a gasoline-electric hybrid version of its ML 450 SUV.

- Toyota—Toyota already produces gas-electric hybrids including the Primus and Camry (assembled at its Georgetown, Kentucky, plant) and has marketed fuel-cell vehicles since 2002. The company recently unveiled a long-distance fuel-cell vehicle that can go 780 kilometers (484 miles) on a charge of 6 kilograms (13 lbs) of hydrogen.[15]

- Nissan—The company has produced an all-electric prototype vehicle called the Mixim that uses lithium-ion batteries, recharges from a wall outlet in under 40 minutes, and sports separate electric motors for the front and rear wheels. The technology for full-scale production is available; however, the costs are too high to bring the Mixim to market.

- Nissan is also partnering with NEC to bring to market smaller laminated lithium-ion cells that can recharge more quickly and store more energy.[16]

- Tata Motors—This India-based car manufacturer has developed the world's least expensive automobile, the Nano—a vehicle that offers the "utility of a car with the affordability of a motorcycle."[17] With a two-cylinder engine and a top speed of 95 kilometers per hour (60 mph), the fuel-efficient gasoline powered vehicle gets 21 kilometers per liter (50 mpg).[18]

The automobile industry is also taking broader action to reduce carbon intensity. By establishing a company sustainability policy, Toyota provides an example of how these programs can be used to establish goals and monitor results. It also demonstrates how cost-effective these programs can be. Recently, Toyota released the first update of its progress toward meeting a 2011 Environmental Action Plan that deals with energy and climate change, air quality, recycling and resource consumption. The company had a target of reducing energy use from a 2001 baseline by 18% (on a per square foot basis). Astonishingly, in 2007 Toyota announced that it had already exceeded the goal and achieved a reduction of 23%.[19] Toyota expends over $100 million per year for energy, emitting 1.4 million metric tons of CO_2 annually.[20] The company has invested over $6.5 million for energy efficiency improvements and achieved over $11 million in avoided costs.[21] The reductions have provided a cumulative savings of 2.4 million therms of natural gas and 41 million kWh of electricity—resulting in a 73% decrease in CO_2 emissions.[22] They have also reduced water consumption from 0.98 kgal/vehicle to 0.80 kgal/vehicle.[23] Using a third-party financier, they plan to install a 2.28-MW solar electric power system at a parts center in Ontario, California.

UTILITY COMPANIES

Utility companies find themselves in a quandary as to how to increase electrical production without increasing carbon emissions. According to Basin Electric's CEO and General Manager Ron Harper, "The electric industry is going to build a significant number of power plants, many of them coal-based, in the coming years to meet our nation's growing electrical demand... The questions of what to do with the carbon dioxide produced by the plants is casting a shadow over their viability, and the federal government should undertake an aggressive strategy to mitigate the risk of a carbon condensed future."[24] It is interesting that an industry with a past mantra of "see no carbon, hear no carbon, speak no carbon" is now more progressive in its desire to find mitigation strategies. This is due in part to the fact that "clean coal" technologies are costly to implement and have not proved themselves to be scalable. Regardless, many public utilities are adapting to the new reality that carbon management programs are needed.

When building coal-fired plants, utilities often choose between technologies such as an integrated gasification combined cycle (IGCC) or supercritical pulverized coal (SPC). While the costs for an IGCC plant are higher, adding carbon capture not only adds additional costs but may also reduce plant efficiency. Carbon capture technologies for SPC are not yet commercially available. Since there are few real-world examples, the added costs and economic unknowns create business risks that are difficult to estimate. An alternative to SPC that would enable carbon capture and storage (CCS) might be to use oxygen in the combustion process rather than air. However, the technology needs to be demonstrated and made commercially available. In the U.S., there are few federal incentives to support commercial implementation of CCS.

Utility companies are seeing a growth in demand for clean energy sources that use proven, readily available technologies. Clean energy sources result from investments in wind, solar, biomass and improvements in energy efficiency; they effectively reduce CO_2e. Another reason for the growth in clean energy demand, besides consumer trends, is progressive state policies and changing legislation. In 2003, 10 states required clean energy to constitute 30% of their energy portfolios within 5 to 15 years.[25] Today there are 25 states with such laws. In 2005, the Atlantic County Utilities Authority in New Jersey dedicated its hybrid solar-wind power plant that produces enough power for 3,800 homes, displacing the need for an estimated 24,000 barrels of oil annually.

With legislative change and the increasing availability of incentives, areas in the U.S. will soon see the demand for green energy outpacing the supply. In New England, the shortfall may exceed 1,500 megawatts.[26] As more states pass renewable energy regulations and place greater demands on electrical system infrastructure, green energy shortfalls will increase. According to a report by the National Renewable Energy Lab, by 2010, clean energy demand will outpace generation by at least 37%.[27] When regulated utilities fail to meet legislated clean energy requirements they often face the potential of stiff penalties.

Another problem associated with the delivery of clean energy is the lack of existing transmission capacity. Existing electrical transmission capacity is designed to bring power from central power plants to the locations where it is needed, rather than from decentralized production facilities in close proximity to major electrical loads. Alternative energy systems can have the advantage of being placed closer to locations where electrical production is needed. One example is rooftop solar—photovoltaic (PV) systems that can produce on-site power, using the electricity as needed and selling the excess to their local utility.

To address this problem, utilities are seeking regulatory reforms to allow them to increase their investments in energy efficiency. A group of eight utilities—Con Edison, Duke Energy, Edison International, Great Plains Energy, Pepco Holdings, PNM Resources, Sierra Pacific Resources, and Xcel Energy—that serve over 20 million customers in 22 states—are seeking regulatory approval to increase energy efficiency investments by $500 million annually within ten years. This effort will achieve a reduction equivalent to 30 million tons of CO_2e each year.[28] These types of regulatory changes create incentives for utilities to avoid constructing new power generation facilities. According to Jim Rogers, President and CEO of Duke Energy, this "commitment is indicative of the power of energy efficiency in addressing climate change... there has been a chronic underinvestment in energy efficiency in our country. We are determined to fix that by creating the innovative regulatory frameworks that leverage technology to address climate change, reduce power demand and keep our customer's power bills as low as possible."[29] Even without regulatory changes, these utilities plan to invest approximately $1 billion annually from 2008-2010 in energy efficiency. This investment should equal an annual reduction of 5 million tons of carbon emissions and eliminate the need for ten, 500-megawatt power plants.[30] According to Jeff Sterba, President and CEO of PMN Resources, "Energy efficiency and demand-side management hold immense promise for reducing carbon emissions and moderating the impact of rising prices on customer bills... we believe that utilities, state policy makers and others can unlock the enormous potential of energy efficiency if we work proactively together to encourage this kind of investment."[31]

Making these types of changes means that utilities need to rethink their products and services, morphing their organizations and updating their corporate philosophies. Some utilities are getting creative. In 2005, Seattle City Light became "the first major U.S. electric utility to achieve net zero emissions of greenhouse gases."[32]

SERVICE INDUSTRIES

Corporations are exploring solutions that are appropriate to their operations and facility support locations. The efforts of UPS to reduce carbon emissions provide an example of the difficulties that a growing company experiences. The company has 94,500 ground fleet vehicles.[33] UPS used computer systems to optimize delivery routes from Worldport, its air cargo sorting facility in Louisville, Kentucky. This system has eliminated 45.9 million kilometers (28.5 million miles) of driving to date. The classic brown delivery trucks UPS uses for local deliveries guzzle about 3.4 to 4.2 km/l (8 to 10 mpg) of gasoline. Thus, the savings incurred from optimizing delivery routes is roughly 12.1 million liters (3.2 million gallons) of fuel. Despite this initiative and the company's goal of decreasing total CO_2 emissions produced by operations, actual performance from 2005 to 2006 resulted in a 6.6% increase in emissions. The increase was partly a result of growth in ground and air delivery volume.[34]

In addition to its gasoline-powered vehicles, UPS operates the industry's largest alternative fuel and low-emissions vehicle fleet. In an effort to improve its performance the company is exploring hydraulic hybrid and hybrid-electric vehicles.[35] The hydraulic hybrid uses no batteries or electric motors, and instead mates the truck's diesel engine with a hydraulic propulsion system, consisting of a series of hydraulic storage tanks and pumps.[36] This system replaces the conventional drivetrain and offers benefits similar to other gas/diesel-electric hybrids such as regenerative braking. The system is capable of increasing the fuel economy and reducing CO_2 emissions by 40%.[37] UPS has recently secured leases for 42 three-wheeled Xebra Trucks (an electric utility vehicle) as part of a pilot program in Petaluma, California, to reduce emissions, increase profits and reduce transportation costs.[38] The company has also committed to the purchase of 167 delivery vehicles that use compressed natural gas, in addition to approximately 800 such vehicles they already use in the U.S. The new vehicles will be deployed in Dallas, Atlanta and four locations in California.

Using a $515,000 federal grant, 366 vehicles at the UPS Worldport facility that are used to lift cargo containers, reposition aircraft and carry jet fuel to planes are being switched to operate using a blend of 5% biodiesel and 95% conventional fuel.[39] According to UPS spokesperson Mike Mangeot, the Worldport location was chosen for this pilot project because "it's our largest off-road diesel operation and it is the place where we felt we could have an impact with the diesel initiative."[40] In addition, the company is ordering 139 propane vehicles to add to its fleet of nearly 600 propane trucks already in use in Canada and Mexico.[41]

GREEN BANKING AND FINANCIAL SERVICES

Financial institutions are marketing green consumer programs. Rabobank (a Netherlands-based bank with outlets in California) introduced the Climate Credit Card. The card pays an amount based on the impact of the purchase on greenhouse gas emissions to the World Wide Fund for Nature.[42] The funds are used for environment-friendly projects. If you think this is small change... the bank has captured over 1.1 million customers since its inception. Crittenden Bank, Vermont's largest, offers a program that allows its customers to defray interest in deposit accounts, applying the proceeds to lend the funds for community development and "green" projects.[43] Tucson Federal Credit Union has opened "bike friendly" drive-though lanes at its branch locations.[44] The Banner Bank Building in downtown Boise, Idaho, is an 11-story office building (17,465 m² or 188,000 ft²) that uses geothermal heating and 60% less energy than a comparable building of its size and type.[45] This LEED Platinum building also uses urban storm water for toilets and urinals. Parking fees at the site are graduated based on vehicle fuel efficiency.

Bank of America—the second largest bank in the U.S.—supports the idea of green banking. According to CEO Kenneth Lewis, "We have a tremendous opportunity to support our customer's efforts to build an environmentally sustainable economy—through innovative home and office construction, new manufacturing technology, changes in transportation, and new ways to supply energy."[46] Initiatives at the Bank of America include:

- Allocating $18 billion in commercial green lending and financing for projects.

- Investing $2 billion in consumer programs and efforts to reduce the environmental impacts and greenhouse gas emissions of its operations.

- Changing loan criteria for commercial underwriting to include whether the business creates "sustainable products, services and technologies."[47]

In December 2005, HSBC Bank USA, N.A. became the world's first major bank to become carbon neutral.[48] To meet this goal, the company pledged to reduce environmental impacts with regard to wastes, water, carbon dioxide emissions and business travel.[49] This was accomplished in 2005 by implementing a corporate plan to manage and reduce direct emissions and through purchasing offsets. The plan included reducing the carbon intensity of the corporation's electrical use by purchasing electricity from renewable sources when feasible and offsetting the remaining CO_2 by purchasing emission reduction offsets from green projects.[50] During a three-month period in 2005, the company estimated its CO_2 emissions to be 170,000 metric tons. Offsets were purchased from projects in New Zealand, Australia, India and Germany at a cost of $750,000 or $4.43 per metric ton of carbon.[51] The types of projects HSBC supports include wind farms, waste composting, methane capture from agricultural wastes and biomass cogeneration projects. A brief summary of each HSBC project follows:[52]

- The Te Apiti Wind Farm in North Island, New Zealand, has 55 turbines that generate enough electricity to power 45,000 average-sized New Zealand homes per year. This project operates at roughly 45% of maximum capacity, compared to the wind farm average of only 30%. It is also one of the first projects validated by the Gold Standard, one of the most highly regarded certification systems for carbon reduction. In 2005, HSBC purchased carbon offsets equivalent to 125,000 tons of carbon emissions—the same as taking 29,000 cars off the road.

- Organic Waste Composting in Victoria, Australia, produces CO_2 instead of more climate potent CH_4, using aerobic methods. The Australian Greenhouse Friendly scheme issues the carbon reduction credits using rigorous standards, which also ensure against double counting. An independent party (approved by the Australian government) verifies this process. HSBC purchased 15,000 tons of carbon reduction from this project in 2005.

- Sandbeindorf Agricultural Methane Capture in Sandbeindoef, Germany, converts fresh pig and cattle manure into biogas, in lieu of CH_4. Thus, the facility produces heat and electricity while reducing methane emissions by 90%. In addition, the amount of CO_2 used by the project is reduced because the facility produces a large amount of its own heat and electricity. HSBC purchased 14,000 tons of carbon reduction in 2005 from this project, which is equivalent to taking 3,200 cars off the road.

- Vensa Biotex Biomass Cogeneration in Andhra Pradesh, India, turns biomass and agricultural waste into electricity for local industry. This process reduces GHG emissions by reducing the need to generate electricity from non-renewable or fossil energy sources. The surplus electricity generated is fed back into the grid. HSBC purchased 16,000 tons of carbon reduction from this project in 2005, which is equivalent to the amount produced by a Boeing 747 airplane flying roundtrip from New York to Hong Kong.

Community banks are taking their green initiatives directly to their customers. They are using less paper for documentation and filing systems, hosting paper shedding and recycling events, improving their landscaping and sponsoring green activities within their communities. Litchfield Bankcorp in northwest Connecticut has periodic promotions that engage community involvement. Their activities include distributing compact fluorescent lamps and reusable shopping bags. In partnership with a recycling company, they cosponsor events to collect old computers and computer equipment in their parking lot so that the components can be properly recycled and disposed of.

SUPPLY CHAIN ASSESSMENTS

As part of a recently initiated carbon disclosure project (CDP), companies are assessing the emissions of their supply chains (with as many as 50 suppliers) so that investors can be informed of their carbon footprint. The not-for-profit CDP coordinates the data management for requests from over 300 investors and investment groups with $41 trillion of assets under management.[53] The supply chain CDP initiative began with a partnership involving Wal-Mart in September 2007 when the company chose to use the CDP process "to engage its supply chain to report on climate change-relevant information" including effective carbon management strategies.[54] Other companies involved in the CDP supply chain project include Hewlett-Packard, PepsiCo, Tesco, Dell, Schweppes, Imperial Tobacco, Procter & Gamble among others.[55]

THE DARK SIDE OF CARBON CREDITS

While the regulated market for carbon credits is growing and expected to increase to $68 billion by 2010, the unregulated voluntary markets is expected to increase to $4 billion.[56] Due to the rapid expansion of the carbon market, corporations in the U.S. have had a difficult time validating investments in green technologies. Furthermore, a recent investigation by the *Financial Times* highlights the potential for abuse of carbon credits.[57] Findings indicated that:

- There are instances of organizational purchases of credits that failed to yield reductions in carbon emissions.

- Corporations sometimes profit from the sale of carbon credits without implementing reductions.

- Corporations sell carbon credits for efficiency gains that were already implemented.

- Brokers provide services of questionable value.

- Lack of verification or reporting can make it difficult to assess the value of the carbon credits being marketed.

- Companies and individuals can be charged for credits at rates greater than current market value.

- Companies sometimes benefit from selling "carbon credits" in order to reduce their industrial pollution.

Forestry schemes that are used for offsets offer some of the greatest potential for abuse since some of these projects may have moved forward without the financial support of offsets. In addition, offset payments for short-term projects may only postpone deforestation. Opportunities for abuse are lessened when projects are focused on energy efficiency or alternative energy, as these provide a stronger argument that carbon based fuels are actually being replaced. Carbon reductions are also easier to validate. Third party verification of carbon-offset projects reduces the potential for abuse.

SUMMARY

Corporate carbon reduction programs can be successful. Businesses that have documented sustainability programs are more likely to have policies in place that will support long-term carbon reduction efforts. However, these sustainability programs must extend beyond building management programs and simply providing offsets for corporate travel. To be effective, they must focus on a corporation's core products and manufacturing processes. Energy saving and alternative energy improvements are often the most effective ways to reduce carbon emissions in corporate operations. They provide life cycle cost reductions and are more likely to provide the opportunity to validate carbon reductions. Since carbon markets are expanding, corporations have had difficulties validating investments in green technologies. Third-party verification of carbon reductions over an extended period of time is important.

Endnotes

1. Gale, S. (2007, September 14). Carbon reduction wins mega brand attention at conference. *GreenBiz*. http://www.greenbiz.com/news/news_third.cfm?NewsID=35906, accessed 26 September 2007.
2. Muraya, N. (2008). Austin climate protection plan "possibly the most aggressive City greenhouse-gas reduction plan." *Energy Engineering*. 105 (2). pgs. 43-44.
3. Lubin, J. (2008, August 17). Environment attracts corporate interest. *The Courier-Journal*. p. D3.
4. *Ibid.*
5. Muraya, N. (2008). Austin climate protection plan "possibly the most aggressive city greenhouse-gas reduction plan." *Energy Engineering*. 105 (2).
6. *Ibid.*, p. 41.
7. *Ibid.*, p. 43.
8. *Ibid.*
9. Healey, J., Woodyard, C. and Carty, S. (2007, September 18). Alternative power sources for autos drive into spotlight. *USA Today*. p. 3B.

10. Naughton, K. (2007, December 31). The man who revived the electric car. *Newsweek*. p. 74.

11. Bob Lutz also was also the GM executive responsible for the Dodge Viper and the 1,000 HP Cadillac Sixteen.

12. Chevrolet. *Concept Chevy Volt*. http://www.chevrolet.com/electriccar/, accessed 8 January 2008. Details regarding the configuration of the hybrid propulsion system can be found at: http://www.chevrolet.com/pop/electriccar/2007/process_en.jsp

13. *Ibid.*

14. O'Dell, J. (2006, September 25). Honda unveils 'super-clean diesel engine'. *The Los Angeles Times*.

15. Takahashi, T. (2007, September 29). Toyota demonstrates long-distance fuel-cell car. *The Courier-Journal*. p. D4.

16. *Ibid.*

17. Rabinowitz, G. (2008, January 11). Modern Model T arrives. *The Courier-Journal*. p. D1.

18. *Ibid*

19. Environmental Leader (2007, December 7). *Toyota releases Environmental Report, plans 2.2 MW solar power system.* http://www.environmentalleader.com/2007/12/07/toyota-releases-environmental-report-plans-22-mw-solar-power-system/, accessed 3 May 2008.

20. *Ibid.*

21. *Ibid.*

22. *Ibid.*

23. *Ibid.*

24. Basin Electric Power Cooperative (2007, September 17). *Harper weights in on capture and storage.* http://www.basinelectric.com/NewsCenter/News/Briefs/Harper_weighs_in_on_.html, accessed 17 October 2007.

25. Davidson, P. (2007, October 4). Clean energy can't meet growing demand. *USA Today*. http://www.usatoday.com/money/industries/energy/environment/2007-10-03-clean-energy_N.htm?csp=34, accessed 5 October 2008.

26. *Ibid.*

27. *Ibid.*

28. Duke Energy (2007, September 27). *Eight utilities seek to increase energy efficiency investment $500 million annually.* http://www.duke-energy.com/news/releases/2007092701.asp, accessed 5 October 2007.

29. *Ibid.*

30. *Ibid.*

31. *Ibid.*

32. Geri (2006, October 23). *U.S. mayors embrace Kyoto Protocol.* http://www.goodnewsnetwork.org/earth/general/cities-embrace-kyoto.html, accessed 27 April 2008.

33. Korzeniewski, J. (2007, August 21). *UPS has goals for reducing fuel use and emissions.* http://www.autobloggreen.com/2007/08/21/ups-has-goals-for-reducing-their-fuel-use-and-emissions-how-ar/, accessed 18 January 2008.

34. *Ibid.*

35. Wolfe, B. (2007, October 9). UPS converting to biodiesel for pilot project. *The Courier-Journal*. p. D1.

36. Nunez, A. (2006, June 26). *EPA unveils hydraulic hybrid UPS delivery truck.* http://www.autoblog.com/2006/06/26/epa-unveils-hydraulic-hybrid-ups-delivery-truck/, accessed 8 January 2008.

37. *Ibid.*

38. Triplepundit (2008, January 15). *UPS launches a small zero emissions fleet.* http://

www.triplepundit.com/pages/ups-launches-a-small-zero-emis-002848.php, accessed 18 January 2008.

39. *Ibid.*
40. *Ibid.*
41. *Ibid.*
42. Dunn, C. (2006, October 2). *Dutch bank introduces 'Climate Credit Card'.* http://www.treehugger.com/files/2006/10/dutch_bank_intr.php, accessed 20 October 2007.
43. Sommer, S. (2007, October 5). Green banking. *Gonzo Banker.* http://www.gonzo-banker.com/article.aspx?Article=350, accessed 20 October 2007.
44. *Ibid.*
45. The Christensen Corporation. *Banner bank building.* http://www.hdrinc.com/13/38/1/default.aspx?projectID=406, accessed 20 October 2007.
46. Bank of America (2007, March 6). *Bank of America commits $20 billion to green lending.* http://blogs.business2.com/greenwombat/2007/03/bank_america_co.html, accessed 19 October 2007.
47. *Ibid.*
48. HSBC (2007). *HSBC Bank Canada in the Community.* p. 12.
49. *Ibid.*
50. HSBC (2007). *Reducing emissions.* http://www.hsbccommittochange.com/environment/hsbc-case-studies/carbon-dioxide/reducing-emissions/index.aspx, accessed 28 May 2008.
51. *Ibid.*
52. *Ibid.*
53. Bloomburg News (2008, January 21). Companies join effort to curb emissions. *USA Today.* p. 5B.
54. Carbon Disclosure Project. http://www.cdproject.net/sc_home.asp, accessed 21 January 2008.
55. Bloomburg News (2008, January 21). Companies join effort to curb emissions. *USA Today.* p. 5B.
56. Harvey, F. and Fidler, S. (2007, April 25). Industry caught in carbon 'smoke-screen.' *Financial Times.* http://www.ft.com/cms/s/0/48e334ce-f355-11db-9845-000b5df10621.html, accessed 1 October 2007.
57. *Ibid.*

Chapter 10

Industrial and Manufacturing Carbon Reduction Technologies

"Business organizations play an increasingly important role in the world. Companies that win the public's confidence and trust are open, visible, engaging and create business value while delivering benefits to society and the environment."

William Ford, Chairman and CEO of Ford Motor Company, (2002)[1]

IMPACT OF INDUSTRIAL CARBON EMISSIONS

Industry accounts for 50% of the world's energy use.[2] United States industry alone consumes 15% (or 34 QBtus) of global energy production, and accounts for nearly one-third of U.S. carbon emissions.[3] Interestingly, U.S. industry represents 38% of the global opportunity for reducing carbon through energy efficiency.[4]

Energy efficiency is the most cost-effective carbon reduction option. Since U.S. energy use is more than any of the other Group of Seven (G7) nations, large opportunities exist for energy reduction, GHG and CO_2 emissions reduction, and fuel flexibility.[5] Overall, the U.S. industrial sector represents 37% of the country's natural gas demand and 29% of total electrical demand.[6] A goal established by the G. W. Bush administration was to reduce U.S. carbon intensity 18% by 2012. The U.S. Department of Energy's Industrial Technologies Program (ITP) has set a goal to reduce industrial energy intensity 25% by 2017. With the exception of incentives offered in the 2005 Federal Energy Policy Act to improve energy efficiency, no federal funding was actually provided to achieve either of these goals.

In order to accomplish these goals, current U.S. industrial energy use must be considered by fuel type—29.8% petroleum, 24.8% natural gas, 23.8% electricity losses, 10.8% electricity, and 4.4% renewables. Any GHG and CO_2 emissions reduction targets would use such data to establish realistic objectives. Many U.S. industries, including Dow, DuPont and 3M are aggressively reducing their CO_2 emissions by implementing profitable energy-efficiency improvements. Dow has saved billions of dollars since 1995 by reducing the energy intensity of its products while successfully reducing its CO_2 emissions by 20%.[7]

The rapid economic growth of emerging countries such as China and India—together with continued, moderate growth in today's economies—could have serious long-term consequences for global energy consumption and carbon emissions. The IPCC's Mitigation of Climate Change Report states:

> "Development policies not explicitly targeting GHG emissions can influence these emissions in a major way. For example, six developing countries—Brazil, China, India, Mexico, South Africa, and Turkey—have avoided, through development policy decisions, 300 million tons a year of carbon emissions over the past three decades. Many of these efforts were motivated by common drivers, such as economic development and poverty alleviation, energy security, and local environmental protection. A case in point: in China, growth in GHG emissions has been slowed to almost half the economic growth rate over the past two decades through economic reform, energy efficiency improvements, switching from coal to natural gas, renewable energy development, afforestation, and slowing population growth. In India, key factors in GHG emissions reduction have been economic restructuring, local environmental protection, and technological change, mediated through economic reform, enforcement of clean air laws by the nation's highest court, renewable energy incentives and development programs funded by the national government and foreign donors."[8]

The projections demonstrate that if countries adopt a "business-as-usual" approach, the potential result could more than double global carbon emissions by 2050 (Stern Review 2006). An increase of this magnitude would have serious long-term implications for climate change. To be effective, the industrial sector must align its goals and strategies to stabilize atmospheric carbon concentrations.

Pricewaterhouse-Coopers (PWC) published a relevant report from a global perspective entitled "The World in 2050: Implications of Global Growth for Carbon Emissions and Climate Change Policy." This report was prepared in advance of the *Stern Review Report* and outlined a "Green Growth Plus" strategy that allows for healthy economic growth while controlling carbon emissions. The report considers six possible scenarios but focuses most attention on two key possibilities:

1) A baseline scenario in which energy efficiency improves in line with trends of the past 25 years, with no change in fuel mix by country. This "business-as-usual" scenario is a benchmark against which the need for change can be evaluated, rather than a forecast of the most likely outcome; and

2) A scenario called "Green Growth + CCS," which incorporates possible emission reductions due to a greener fuel mix, annual energy efficiency gains over and above the historic trend, and widespread use of CCS technologies. This "Green Growth Plus" strategy stabilizes atmospheric CO_2 concentrations by 2050 at acceptable levels.[9]

THE METRICS OF CARBON REDUCTION

GHG emissions measurement calculates baselines of energy consumption or GHG emissions that use industry-specific metrics.[10] Businesses, governments, and analysts that track trends in GHG emissions and energy consumption commonly use sectorial and industry-specific metrics, also called "indicators." These metrics are designed to measure changes in CO_2 intensity or energy efficiency, independent of economic growth or growth in production. Indicators typically use either an economic or physical value for the denominator. For example, the energy intensity of cement production can be measured as energy use per dollar of value added by the cement industry (economic metric) or energy use per ton of cement produced (physical metric). Economic metrics are typically used when aggregating across heterogeneous entities that do not produce comparable products (e.g., the entire manufacturing sector). Physical metrics are typically used to compare entities that have similar production outputs.

According to the IPPC 2007 Mitigation of Climate Change report, carbon intensity is defined as CO_2 emissions per unit of total primary energy supply (TPES). Energy intensity is defined as TPES per unit of gross domestic product. From a baseline scenario perspective, the energy-related emissions are a function of population growth, gross domestic product per person, changes in energy intensity and the carbon intensity of energy consumption. In the development of industry-specific metrics, various methodologies, benchmarking programs, inventories, protocols, and registries that use intensity-based metrics to track trends in energy use or GHG emissions are identified. Then the most appropriate or historically relevant indicators can be selected. The next step in developing industry metrics is to assess the availability of data needed to determine metric development priorities. Accounting for actions to reduce GHG emissions can be performed on a project-by-project basis or an entity basis.

Benchmarking is widely used in industry, but it must be carefully designed to comply with laws ensuring fair competition. The "top down" or computational general equilibrium (CGE) and "bottom up" or technology-rich economic models used for climate change are complementary when applied to the industrial sector of the economy. According to the Pew Center for Global Climate Change, industries responsible for a predominant share of greenhouse gas emissions, that bear proportionate burden, include coal mines, oil refineries, natural gas importers, electricity producers, cement works, chemical facilities, paper and pulp mills, steel and other metal companies, and glass manufacturers. However, mechanisms exist to compensate these sectors, such as revenues from emission permits or taxes, and offering transition assistance to affected communities and workers.[11] Establishing project-related baselines for mitigation efforts has been widely discussed in the context of the "flexible mechanisms" of the Kyoto Protocol, and the United Nations' Framework Convention on Climate Change—Joint Implementation (JI) and the Clean Development Mechanism (CDM). Issues regarding entity-specific baselines which can be used by companies, municipalities, and organizations have been explored in the context of baseline protection, emission trading, credit for early action initiatives, and climate change registries.

Based on the research completed, LBNL recommends the development of a GHG Intensity Index as a new metric for reporting and tracking GHG emissions trends. LBNL developed a baseline typology and assessed the complexity and robustness of each type of baseline vis-à-vis potential future emissions limits or emissions trading schemes. LBNL conducted three case studies in order to explore issues related to both industry-specific metrics and baselines. The case studies demonstrated numerous issues related to the use of metrics. LBNL recommended that industry-specific metrics be disaggregated to a certain degree, depending upon the specific sector and data availability, in order to fully comprehend the energy or GHG emissions trends at a given industrial facility.

MODELING SYSTEMS

The EIA has proposed a National Energy Modeling System (NEMS) to evaluate the effects of the Kyoto Protocol in relation to the energy market results of carbon reduction and sensitivity cases. The first set of cases examines the impacts of six carbon emissions reduction targets, relative to a reference case without the Kyoto Protocol. The remaining cases examine the sensitivity of those results to variations in key assumptions—the macro-economic growth rate, the rate of technological progress, and the role of nuclear power. With regard to the carbon reductions under the Kyoto Protocol, many U.S. state and local governments have committed to reducing GHG emissions to 7% below 1990 levels in the period 2008 to 2012, despite the failure of the central government to ratify the treaty. Reductions in energy-related carbon emissions that the U.S. must achieve depends on many variables such as the level of emission offsets credited for sinks, reductions in other GHGs, international permit trading, Joint Implementation, and CDM of the Kyoto Protocol.[12]

The NEMS modeling approach assumes perfect foresight of carbon prices for capacity planning in the electricity industry. In this context, perfect foresight refers to an accurate prediction of future carbon prices during the planning process. An algorithm is used to estimate carbon prices in which anticipated and realized carbon prices are approximately the same while ensuring that the carbon prices clear the carbon permit markets. In the end-use demand sectors (e.g., the industrial sector), foresight does not typically have a material influence on energy equipment decisions. Instead, such decisions are often based on current market prices at the time of the decision.

Other modeling techniques include a combination of the "bottom-up" and the "top-down" approaches.[13] The Industrial Energy End Use Simulation Model (ENUSIM) represents a technology-based, bottom-up model that was used to partially evaluate the effect of the climate change levy (CCL) on selected industrial sectors in the UK. ENUSIM is designed to model the uptake or retrofit of energy saving or fuel switching technologies in selected industrial sectors, considering economic factors and investments in new technology. The industry-specific abatement cost curves contained in ENUSIM do not, however, include CHP facilities. The "bottom-up" models are technology-rich with the cost of improvements ranging up to $50 per ton of CO_2 avoided.

Thus, the estimated impact of the selected policies and measures on CHP as well as the resulting emissions savings were estimated using the Multi-Sector Dynamic Model Energy-Environment-Economy (MDME3). The MDME3 is a "top-down" model of the UK economy, with fully integrated energy-environment sub-models. These models are also known as computational general equilibrium (CGE) models and value carbon avoidance from $25 to $200 per ton. Likewise, the impact of the climate change agreements (CCAs) on carbon emissions were calculated 'off model' using ENSUIM data. Every effort was made to ensure consistency between the different modeling approaches. Regardless, care should be exercised when aggregating the estimates from the different policy measures.

SPECIAL APPLICATIONS FOR THE INDUSTRIAL SECTOR

A 2007 Special Report of the IPCC, "Carbon Dioxide Capture and Storage," provides valuable information for the industrial sector. This report describes the sources, capture, transport, and storage of CO_2. It discusses the costs, economic potential, and societal issues associated with the technologies (including public perception and regulatory aspects). Storage options evaluated in the report include: geological storage, ocean storage, and mineral carbonation. The report places CO_2 capture and storage in the context of other climate change mitigation options. It also includes summary information for policy-makers and overviews of technologies as approved by the governments represented in the IPCC. Altogether, the report provided invaluable information for researchers in environmental science, geology, engineering, and the oil and gas sector.[14]

A Dutch researcher, Kay Damen, investigated the technical possibilities, costs, and risks of CO_2 capture, transport, and underground storage. Damen found that electricity generation has the greatest potential to avoid CO_2 emissions.[15] The research estimates that 15 million tons of CO_2 per year can be avoided by 2020. Such an achievement would require the implementation of CO_2 capturing technologies for new coal-fired power stations. In 2050, the reduction potential is estimated to increase from 60 to 84 million tons of CO_2 per year, for a scenario in which electricity production is doubled. By capturing CO_2 in industrial processes a further 16 million tons of CO_2 per year can be avoided. For the production of hydrogen in the transportation sector, Damen investigated the thermodynamic performance and costs of decentralized membrane reformers. This new technology makes it possible to capture CO_2 with relatively low costs.

The steel-manufacturing sector provides an example of an industry that is struggling to reduce its carbon footprint. Indeed, the International Iron and Steel Institute (IISI) in London acknowledged that the steel industry accounts for the greatest amount of CO_2 emissions due to its sheer size. Improving energy efficiency in China, which accounts for approximately half of the steel sector's CO_2 emissions, and Russia can make a significant impact.[16] In the long term, the steel industry must boost research and development spending to find ways to either capture and store CO_2 or find new technological solutions for its production processes. In April 2008, the IISI launched the second phase of its "global sectorial approach to climate change," establishing a methodology for collecting and measuring sectorial CO_2 emissions data. Hopefully, this approach will enable the organization to gather relevant data and further expand R&D efforts.[17]

INDUSTRIAL CARBON REDUCTION TECHNOLOGIES

In the report "Carbon Dioxide Emission Reduction Technologies and Measures in U.S. Industrial Sector," researchers identified energy efficient technologies currently used by industries in the United States. The report evaluated the potential energy savings of the various technologies in terms of energy, environment, and economy (3Es), noting that the industrial sector consumes about 33% of all energy used in the United States.[18] Since the sector uses such large amounts of energy, there is potential to reduce carbon emissions by improving efficiency. The U.S. DOE's Industrial Technology Program (ITP) focuses on the eight most energy intensive industries—Aluminum, Chemicals, Forest Products, Glass, Metal Casting, Mining, Petroleum Refining, and Steel—for development of energy efficiency technologies and processes. Of these, five industries were selected for further study: Steel, Petroleum Refining, Petro-Chemical, Cement, and Paper & Pulp. The following descriptions provide a summary of CO_2 reduction potential, as a function of energy savings, for each of the five industries:

a) Steel Industry—The energy intensity of the U.S. iron and steel industry has improved by 28% since 1990. This change is pri-

marily due to energy efficiency improvements. The steel industry uses energy to supply heat and power plant operations, as
well as a raw material for the production of blast furnace coke.
Effective technologies that reduce both energy consumption
and CO_2 emissions focus on combustion systems (e.g., furnaces,
ovens, boilers, and direct-fired operations), fuel switching (e.g.,
natural gas for fuel oil), thermal systems (steam), waste heat recovery, cogeneration, motor systems (e.g. reduced compressed
air use), and more efficient electrical equipment, to replace existing hydraulic or pneumatic equipment. For reporting purposes, carbon equivalence is obtained by using CO_2 emission
coefficients of 0.165 TC/MWH for electricity, 14.47 MTC/Quad
Btus for natural gas, and 19.95 MTC/Quad Btus for fuel oil.[19]
Specific recommendations include: furnace insulation, pulverized coal injection, injection of natural gas, sensible heat recovery (suppressed combustion), coke-less iron making, thin-slab
casting, steam trap maintenance, pre-heat combustion air with
waste heat recovery, and simple cycle or combined cycle cogeneration.

b) Petroleum Refining Industry—Refinery gas, natural gas, and
    coke are the main fuels used by the petroleum refining industry.
    Natural gas and electricity represent the largest share of pur-
    chased fuel in the refineries. However, due to the large amounts
    of self-produced electricity in this industry only a small quantity
    of purchased electrical power is required. Most petroleum re-
    fineries are capable of improving their overall energy efficiency
    by as much as 20%. Such improvements are available from fired
    heaters, process optimization, heat exchangers, and motors.
    Examples of efficiency improvements include: boiler insulation,
    boiler maintenance, steam distribution system insulation, steam
    trap maintenance, recovery of condensate, and correctly sizing
    pumps for electricity savings. CO_2 emission reductions ranging
    from 5 to 15% can be achieved through fuel switching and by
    installing equipment to utilize waste fuels, such as the recovery
    of flare gas.

c) Petro-chemical Industry—This industry is the second largest
    consumer of natural gas and the largest consumer of liquefied

petroleum gas (LPG). 25 MTC of CO_2 emissions have been associated with this industry, including industrial organics. Some of the key chemical products are ethylene, propylene, butadiene, and methanol (e.g., steam cracking derivatives). A 5 to 15% CO_2 emission reduction can be achieved by optimizing the efficiency of hot air furnaces, repairing or replacing steam traps, repairing leaks in lines and valves (e.g., steam leaks), using hot process fluids for preheat, and using waste heat to produce steam for turbine generators (e.g., topping and bottoming cycles).

d) Cement Industry—The primary types of cement products manufactured and used in the U.S. are Portland cement— nearly 90% of total cement usage—with masonry cement comprising the remainder. Indeed, the largest CO_2 emissions are associated with clinker production, which occurs in the kiln in the manufacture of Portland cement. The combustion of fossil fuels and calcinations of limestone contribute the majority of CO_2 emissions in this industry. Some of the effective energy and CO_2 emission reduction measures include: kiln combustion system improvements and shell heat loss reduction, installation of cogeneration systems for electrical power, and use of high efficiency motors with variable speed drives. As much as 20 to 25% reduction in energy use is achievable through the implementation of these technologies.

e) Pulp and Paper Industry—The industry manufactures products such as pulp from wood or wastepaper, paper from wood or fiber pulp and paperboard products from wood and fiber pulp. The main pulping processes involve mechanical, chemical, and semi-chemical systems. Steam and electricity are the major energy fuels for these processes. Energy and CO_2 emission reductions as great as 30% can be achieved through the use of efficient equipment, maintenance, and cogeneration (e.g., combined cycle gas turbine) technologies.

UTILITY CASE STUDIES

Descriptions of two utility case studies are provided in the following section. The first is an example of a U.S. utility, Duke Energy Corporation, and the second involves Dalkia Utility Services in partnership with Avencia Limited in the UK.

Headquartered in Charlotte, North Carolina, Duke Energy Corporation is among the largest electric power companies in the United States, supplying and delivering energy to approximately 3.9 million customers. The company has nearly 37,000 MW of electric generating capacity in the Carolinas and Midwest, as well as natural gas distribution services in Ohio and Kentucky. In addition, Duke Energy has more than 4,000 MW of electric generation in Latin America. Duke Energy's annual CO_2 emissions from generation operations are about 107 million tons. In response, Duke Energy has established a goal of avoiding or offsetting 10 million tons of CO_2 by 2015. It has committed an annual investment of $3 million to fund GHG-reducing projects.

Duke Energy thinks of energy efficiency as the "fifth fuel" in their business. Energy efficiency joins coal, natural gas, nuclear, and renewables as a critical resource needed to serve the growing energy needs of the communities served by Duke Energy. The utility is committed to working with its state regulators in developing energy efficiency programs that both save its customers money and improve the environment. It has committed to retiring one MW of older, less efficient coal-fired generation for each MW it saves through its energy efficiency programs. Duke Energy recently filed a tariff request in North Carolina, asking regulators to approve its "Save-A-Watt" energy efficiency program. This program will cost customers approximately 10% less than the cost of building and operating new power plants. Duke Energy plans on filing similar tariff requests in the other four states it serves.

The company is leading the national effort, collaborating with the U.S. DOE, the U.S. Environmental Protection Agency (USEPA), state regulators and other utilities to produce the National Action Plan for Energy Efficiency. This effort is co-chaired by Duke Energy's CEO and President, Jim Rogers. Duke Energy reports its GHG emissions through the U.S. DOE's voluntary 1605b reporting process, and more recently through the U.S. EPA's Climate Partners program. The company also participates in the Carbon Disclosure Project.

Duke Energy joins in numerous forums for information exchange, research, and dialog on topics related to global climate change, including the Alliance to Save Energy, Edison Electric Institute, Electric Power Research Institute, Global Environmental Management Initiative, U.S. EPA's Climate Leaders, Pew Center on Global Climate Change, and

Resources for the Future. Duke Energy is also a member of the United States Climate Action Partnership, a group of corporations and environmental NGOs (non-governmental organizations) that has called for economy-wide legislation to control U.S. GHG emissions.

Duke Energy sees that low or zero-emission technologies will play an increasingly larger role in meeting future energy demand in the U.S. Therefore, the company is notably active in bringing more renewable energy to the U.S. market. The company provides opportunities for customers in several states to support renewable energy development, including the recently approved GoGreen tariff in Ohio. Similar company tariffs, available in Indiana and North Carolina, provide customers the opportunity to purchase renewables for a portion of their energy needs. Duke Energy Carolinas has recently issued a "Request for Proposals (RFPs)," seeking bids for power generated from renewable energy sources, including solar, wind, hydro, and biomass, among others. In Indiana, Duke has contracted to purchase power from a 100 MW wind farm that is being developed. In northern Kentucky, the company offers rebates for commercial, governmental, and institutional customers who make energy saving improvements to their facilities.

The partnership between Avecia Limited and Dalkia Utilities Services offers an example of a project whose goals were to provide a stable energy supply, significant cost savings, and enhanced environmental improvements through implementation of a combined heat and power (CHP) plant. Avecia (now Kemfine) wanted to concentrate resources on core business activities and subtract energy services, benefitting from the cost-savings and environmental improvements that new CHP plants can offer without using their own capital. At Avecia Limited's Grangemouth site, the company manufactures textile dyes, industrial and specialist colors, fine chemicals, agrochemicals, biocides, and pharmaceutical intermediates. At Grangemouth, construction economists from Franklin and Andrews provided commercial management services for the construction of a 130 MW combined heat and power (CHP) station.

Dalkia was able to meet Avecia's criteria and was awarded a 15-year contract energy management agreement. This agreement required a $10 million capital investment for a new energy plant at the Grangemouth site. The new energy facility consists of a 7.5 MW combined cycle CHP plant that provides electricity and steam to the chemical manufacturing process. The contract involved two phases. In phase one, Dalkia Utilities Services operated and maintained the existing energy supply plant for roughly two years, while the new CHP plant was being designed and constructed. In phase two, Dalkai was responsible for providing all energy from the new CHP plant to the Grangemouth site, in the form of metered electricity and steam, through an energy supply agreement. The agreement includes standby electricity, bulk purchase of fuel, warranty of the CHP plant over the 15-year contract, and responsibility for the efficient operation and maintenance of the energy plant. The combined cycle CHP plant is comprised of a 4.5 MW gas turbine, a 3 MW steam turbine, and a 30-tons/hour waste heat boiler to provide optimum availability and flexibility in meeting the site's varying energy needs. A new powerhouse was constructed for all the equipment, including an energy monitoring and control system (EMCS). The implementation of the CHP strategy has significantly improved the efficiency with which steam and energy are supplied to the site. The combined effects of improved energy efficiency and CHP fuel combination provide substantial environmental benefits, including:

- 51% reduction in emissions of CO_2
- 66% reduction in emissions of SO_2
- 63% reduction in emissions of NO_x
- 99% reduction in emissions of particulates

The Grangemouth project provides a model for utility and industry partnerships that are planning to reduce GHG emissions.

MANUFACTURING GOING GREEN

Canon

Manufacturing industries are counting carbon emissions and attempting to reduce their carbon footprint. Many of these programs were initiated in the 1980s as a part of their corporate sustainability programs. In 1998, Canon began a program to reduce their carbon intensity throughout the lifecycle of their products. Over an eight-year period, they developed "energy-saving on demand fixing technology" for their office equipment and home printers—reducing carbon emissions by nearly 7 million tons.[20] This corporate strategy involves analyzing the product life-cycle in terms of the potential to recycle products and components in regard to environmental impact using advanced technologies. This process includes consideration of CO_2 reduction potential for products such as digital cameras and printers.

DuPont

DuPont was an early participant in addressing global climate change, because of its success in addressing the depletion of the ozone layer by developing alternatives to chlorofluorocarbons (CFCs). In the mid-1990s, DuPont adopted an aggressive corporate energy policy that focused on maximizing energy efficiency, lowering the environmental impact of energy conservation and renewing the company's power infrastructure. In order to achieve this, energy teams were established for DuPont's business units and facilities. Each month, the teams calculated detailed metrics for energy used per pound of product, and the steam and electricity usage per building. This information was used to help assess how improvements could be made in the areas of efficient lighting, heating, cooling, compressed air, and cogeneration.

As early as 1991, the company began to inventory its GHG emissions, and identify point source reductions across its global operations. The carbon footprint of DuPont's worldwide operations in 2005 was 13,550,000 metric tons of carbon equivalent. Its targets are to achieve a 65% GHG reduction from 1990 levels by 2010, to hold energy use constant at 1990 levels, and to source 10% of global energy from renewables. Its anticipated benefits include $2 billion saved through increased energy efficiency and $10 to $15 million saved through use of renewables. Achievements have exceeded expectations in most categories. DuPont has reported a 67% reduction in CO_2 emissions since 1990, a 9% reduction in energy use below 1990 levels (while increasing production 35%), and a 3% increase in energy use from renewables such as hydropower. In order to meet its goal of sourcing 10% of its energy requirements with renewable energy, DuPont is purchasing 170 million kWh per year of renewable energy certificates (RECs) generated by projects that produce electricity from biomass and landfill gas.

To help reach its GHG targets in a cost-effective manner, DuPont has participated in several external emissions trading programs, including the Chicago Climate Exchange and the UK Emissions Trading Scheme. DuPont believes that participation in these markets has enabled it to meet a number of goals, such as generating cash flow to defray the cost of emission reductions and educating managers on the value of investing in emission reduction improvements. This proactive stance has also helped the firm develop the tools, information, and strategies necessary to gain a competitive advantage as emission markets evolve.

Among DuPont's climate change initiatives, energy efficiency opportunities have provided the greatest cost savings. The company plans to identify projects with an internal rates of return (IRR) greater than 20% and extend responsibility for its energy saving program to the Asset Productivity Leadership Team. DuPont invested $50 million to retrofit facilities in Texas, Canada, UK, and Singapore, in order to reduce the nitrous oxide (N_2O) emissions from nylon production. Nitrous oxide has 310 times the global warming potency of CO_2. These process changes allowed the company to reduce GHG emissions from its global operations by approximately 55%. Moreover, DuPont has actively explored and invested in the environmental benefits of its products.

International Business Machines or IBM

IBM's Energy and GHG Emissions Management program was established to curb:

- Direct emissions of CO_2 from boilers at IBM sites,
- Indirect emissions of CO_2 as a result of electricity consumption,
- Direct emissions of perfluorocompounds (PFC), and
- The six gases (NF_3, CF_4, C_2F_6, C_3F_8, SF_6, CHF_3) used in semiconductor manufacturing.

As part of the 2002 Climate Leadership Partnership, IBM established two primary goals:

- To achieve an average CO_2 emission reduction equivalent of 4% of the company's annual electricity and fuel use, through further energy conservation actions; and

- To achieve an absolute 10% reduction in perfluorocompound (PFC) emissions from IBM semiconductor manufacturing processes by 2005, using 2000 as the base year.

Based on their emissions reduction analysis, IBM has reduced roughly 58% of PFC emissions, corresponding to over 300,000 million tons of CO_2 equivalent, between the base year 2000 and 2004. In terms of the actual quantity of CO_2 emissions reduced, IBM has achieved approximately 900,000 tons of CO_2 reduction during the five-year period. This figure translates into more than a 6% reduction, almost 50% above the established goals.

In addition, IBM has acquired over 220,000 MW hours of energy from renewable sources worldwide, including energy from CHP plants, wind, biomass, solar, and landfill gas. Thus, over 15 years (1990 to 2004), IBM has conserved 16 billion kWh of electricity or over 16.4% of total electricity use (equivalent to electricity used by 2 million average homes in the U.S. annually); avoided more than 9 million tons of CO_2 emissions (equivalent to emissions from 2 million cars driving 10,000 miles each); and saved about $860 million in expense (equivalent to 73% of 2004 yearly dividends paid to IBM shareholders). For IBM, aggressively pursuing energy conservation and PFC and CO_2 emissions reduction on a long-term basis has been an effective approach. IBM's efforts not only meet the Global Climate Change Challenge, but have also reduced the cost of operations and increased shareholder value.

SUMMARY

The industrial sector accounts for 50% of the world's energy use. Industrial processes and practices represent a significant fraction of the world's total atmospheric carbon emissions. Energy efficiency is the most cost-effective carbon reduction option. The amount of industry-driven carbon emissions is likely to escalate with the continued, rapid development of countries such as China and India, in conjunction with business-as-usual practices in industrial countries. U.S. industry, in particular, has the potential to dramatically impede the anticipated growth in carbon emissions through energy efficiency, energy reduction, fuel flexibility and expanded R&D.

In response to the increased awareness of externalities associated with carbon emissions, industries have evolved. Many now emphasize the importance of carbon reduction in their R&D investments and strategic plans. In addition, new methodologies have surfaced that focus on the development of industry-specific metrics and modeling systems. The purpose is to improve the validity and cost-effectiveness of carbon reduction strategies. The "metrics" establish standard processes for emission measurements and data collection, while modeling systems estimate improvements in carbon emissions or energy efficiency in relation to policies (e.g., the Kyoto Protocol).

In this chapter's examination of industrial practices, special attention is afforded to various technologies that propagate carbon emission reduction. For example, energy efficiency technologies show enormous carbon reduction potential. Industrial process improvements are also receiving more attention in the United States. This chapter outlines progressive strategies and improvements that are implemented by both utility and manufacturing industries. The experiences of both DuPont and IBM provide examples of companies that have made operational changes to achieve CO_2 reductions. A vast array of opportunities exists for the industrial sector to retrofit and reform the current, business-as-usual processes. The industrial sector has made incremental steps towards the implementation and evaluation of more environment-friendly technologies; however, the expansion and sustainability of these changes will largely rely on whether they prove to be cost-effective and applicable to a wide-range of industrial practices.

Endnotes

1. Ford, W. (2002). *Ford Motor Company*.
2. Energy Information Administration (2007). *Annual energy outlook*. www.eia.doe.gov.
3. Energy Information Administration (2007, May). *International energy outlook 2007*. http://www.eia.doe.gov/oiaf/ieo/, accessed 28 May 2008.
4. McKinsey & Company (2007, May). *Curbing global energy demand growth: the energy productivity opportunity*. http://www.mckinsey.com/mgi/publications/Curbing_Global_Energy/index.asp, accessed 28 May 2008.
5. Group Seven (G7) countries include the United States, Canada, France, Germany, Italy, Japan and the United Kingdom. Russia was added later and it was changed to the Group Eight (G8).
6. Kaempf, D. (2007). *U.S. Department of Energy's Industrial Technologies Program*.
7. Miller, P. (2009, March). Saving energy—it starts at home. *National Geographic*. p. 72.
8. Intergovernmental Panel on Climate Change (2007, November). Report on *Mitigation of climate change*. p. 701-702.
9. Hawksworth, J. (2006). *The world in 2050: implications of global growth for carbon emissions and climate change policy*. Pricewaterhouse-Coopers.
10. Lynn, P., Murtishaw, S. and Worrell E. (2003, June). *Evaluation of metrics and baselines for tracking greenhouse gas emissions trends: recommendations for the California climate action registry*. Lawrence Berkeley National Laboratory.
11. See Pew Center on Global Climate Change. *Global warming basics*. www.pewclimate.org/global-warming-basics.
12. Intergovernmental Panel on Climate Change (2007). Report on *Mitigation of climate change*.
13. Pew Center on Global Climate Change (2003, August). *Impact of the climate change*

program on industrial CO_2 emissions.

14. Intergovernmental Panel on Climate Change (2005). *Carbon dioxide capture and storage.* Cambridge: Cambridge University Press.
15. The Netherlands Organization for Scientific Research (2007, April 10). *Prepare CO_2 capture and storage now for greater environmental benefit later.* http://www.nwo.nl/nwohome.nsf/pages/NWOA_6ZKL7V_Eng, accessed 28 May 2008.
16. International Iron and Steel Institute (2008, April 15). *Steel industry launches global CO_2 emissions data collection programme.* http://www.worldsteel.org/index.php?action=newsdetail&id=238, accessed 28 May 2008.
17. *Ibid.*
18. Center for Energy and Environmental Policy (2007). *Carbon dioxide emission reduction technologies and measures in U.S. industrial sector.* University of Delaware.
19. TC = Ton of carbon equivalent, M = Million, Wh = watt hours, and Quad = quadrillion Btus.
20. Canon (2007, September 17). Sustainability is our standard for measuring CO_2 reduction. *Newsweek.* p. E3.

Chapter 11

Organizational Structure and Resources

"Good leaders achieve results: great leaders achieve sustainable results by serving multiple constituencies."

Kevin Cashman, CEO of LeaderSource,
from his book Awakening the Leader Within (2003).[1]

ORGANIZATIONAL STRUCTURES

Any effective or sustainable organization—whether focused on global environmental issues like climate change or engaged in international trade as a multi-national corporation—relies heavily on human resources. In Gayle Avery's *Leadership for Sustainable Futures* seven organizational elements are described within her "Focus on People" chapter:

- Management development
- Ensuring a strong organizational culture
- Making people a priority
- Retaining staff in difficult times
- Creating a skilled workforce
- Managing uncertainty and change
- Fostering cooperative union-management relationships[2]

Specifically, culture can be defined as basic assumptions that people in an organization hold and share about the organization. Those assumptions are implied in their shared feelings, beliefs and values, and embodied in symbols, processes, forms, and some aspects of patterned group behavior. Organizational culture provides clues to the "soft rules" of an enterprise, and is an instrument for managing communication, behavior, and relationships. This is consistent with the advice of a well-known 20th century leadership expert, Margaret Whitley, namely, "If you distance yourself from your people—refusing to cultivate meaningful relationships with them—you are destined to fail."

Rapid technological changes, pressures created by globalization, fierce competition, and increasing resource constraints, make the management of organizational change a critical issue for organizations of all types. The scientific study of organizational change is an interdisciplinary endeavor and it has grown in step with the practical importance of change to organizations. Only a limited part of the science of change literature has directly addressed how to change organizations in ways that would benefit the natural environment or the earth's climate. The process of changing organizations to help address climate change is the focus of this chapter.

Climate change is one of the greatest challenges to international cooperation that the world is currently facing (*Stern Review Report* 2006). Collective action by independent sovereign nations is particularly challenging goal. There is no supranational authority to provide leadership for issues relating to global climate change (such as the World Trade Organization for International Trade). Future cooperation depends on the perception and extent of benefits offered to participating nations and organizations. These entities must recognize significant and sufficient advantages derived from participation in international treaties (e.g., Kyoto Protocol) or other arrangements that share a vision of responsible behavior. Nations and organizations should also consider that, without large-scale involvement, collective action geared at mitigating climate change might fail. For effective climate change mitigation and adaptation, nations and organizations need to restructure themselves.

A typical organizational structure with a consensus-based strategic plan for CO_2 and GHG emissions reduction, should include the following elements:

- Establish an institutional commitment with quantitative goals.
- Build a climate action team with interdisciplinary expertise.
- Conduct CO_2 and GHG inventories and establish a "carbon footprint" regime.
- Develop and implement a climate action plan.
- Establish CO_2 and GHG emissions reduction goals and targets.

INTERNAL AND EXTERNAL RESOURCES

Many countries, regions and cities have adopted initiatives that complement and extend multilateral frameworks. Such frameworks typically negotiate and uphold common national and international goals, which can be implemented at regional and local government levels. In 2005, a comprehensive United Nations Development Program (UNDP) study found that over half of these policy measures flow from national policy choices. Many of the remaining measures are undertaken in cooperation with multilateral organizations.[3] These organizations might include international companies or quasi-governmental organizations, such as the International Atomic Energy Agency (IAEA). In one example, several countries backed the IAEA's initiative by donating funds towards an international nuclear fuel bank. Such a development would ensure that member states of the IAEA could enjoy the environmental and economic benefits of reliable nuclear energy, while in compliance with non-proliferation, security and safeguard obligations.[4]

The resources available to cities are linked to the resources available to their regions and countries. Half of the world population lives in cities and many more travel into cities for work. Urban areas of nation states account for nearly 78% of carbon emissions from human activities.[5] Thus, regional and local government structures, in cooperation with multinational organizations and corporations, have the potential to significantly reduce carbon emissions.

In terms of mobilizing internal and external resources for CO_2 and GHG emissions reduction, an innovative approach has recently been initiated in California, which is one of the top ten economies of the world with a population of 37 million. The state ranks as the second largest emitter of CO_2 in the U.S. The California Public Utilities Commission (CPUC) created the California Institute for Climate Solutions (CICS), whose mission is based on the following fundamental goals:

- "To facilitate mission-oriented, applied and directed research that results in practical technological solutions. This research should support the development of policies that reduce GHG emissions in the electric and natural gas sectors, or otherwise address the impacts of climate change in California.

- To speed the transfer, deployment and commercialization of technologies with the potential to reduce GHG emissions in the electric and natural gas sectors."[6]

The work of the CICS is directed by a strategic plan that focuses on potential areas of research, maximizing consumer benefit, and minimizing unnecessary redundancy. Hence, the strategic plan identifies areas of research and technological innovations that are likely to achieve the greatest GHG reductions at the lowest cost. The CICS is committed to reducing the consequences of climate change on infrastructure, public services and policies. It is reflective of California's vision for a more sustainable future.[7] Furthermore, the CICS has a Governing Board that is responsible for ensuring that its mission is fulfilled. The Strategic Research Committee, which is chosen by the Governing Board, is responsible for three main tasks:

- Developing a strategic plan and providing annual updates;
- Assisting CICS officers in developing short-term and long-term action plans; and
- Reviewing grant proposals recommended by a peer review committee.

The California State Legislature has approved the CICS' budget at $60 million annually for 10 years. However, the CPUC has required that the CICS executive director be responsible for obtaining 100% matching funds over 10 years, to maximize and leverage benefits. This coalition represents an aggressive public-private partnership that involves the government, the business sector, and community at large. Anticipated benefits from this CICS initiative correspond with the State of California's Legislative Assembly Bill 32, which requires California to reduce GHG emissions to 1990 levels by 2020 (25% below the projected business-as-usual levels).[8] In addition, Executive Order S-3-05 (issued by California Governor, Arnold Schwarzenegger) targets an aggressive reduction of 80% of the CO_2 and GHG emissions below 1990 levels by 2050.[9]

Similar innovative and bold steps like California's are being adopted and undertaken by hundreds of other states, metropolitan area, and local entities around the world. Many of these initiatives far exceed the respective national targets. In order to succeed, these public-private partnerships and multi-disciplinary approaches must incorporate internal and external human, financial and technological resources.

Table 11-1. Carbon Dioxide Emissions in 2005 from the Top 20 U.S. States

Rank	State	Million Metric Tons of Carbon Dioxide
1	TX	625.2
2	CA	395.5
3	PA	284.0
4	OH	274.0
5	FL	262.6
6	IL	250.4
7	IN	237.9
8	NY	212.2
9	MI	192.3
10	GA	185.7
11	LA	183.1
12	NC	155.6
13	KY	153.8
14	MO	142.8
15	AL	142.2
16	NJ	133.4
17	VA	130.6
18	TN	125.9
19	WV	114.3
20	WI	112.1

Source: Energy Information Administration: Official Statistics from the United States Government. www.eia.doe.govoiaf/1605/state/state_emission.html, accessed 5 January 2009.

LEADERSHIP AND DECISION-MAKING PROCESSES

Dr. Arun Jhaveri's doctoral dissertation, "Effective Leadership for Sustainable Development in the Public Sector," outlined the intricate relationships between the decision-making processes, public sector policies, effective leadership, and sustainable development principles (environmental stewardship, economic prosperity, and social justice). Together, these factors comprise a model best suited for addressing global climate change and associated CO_2 and GHG emissions reduction activities on a local, regional, national, and international scale.[10] All these intimately connected elements also provide a foundation for the development of strong and sustainable leadership skills. Such leadership is critical to the successful accomplishment of complex climate change targets that are established by organizations and institutions around the world.

The IPCC's Climate Change 2007 Mitigation Report connects decision-making with risks and uncertainties associated with global climate change. Specifically, climate change mitigation policies must account for these risks and uncertainties, which makes decisions on appropriate responses so difficult. Risk refers to cases in which the probability of outcomes and consequences can be ascertained through well-established theories with reliable, complete data. Uncertainty refers to situations in which the appropriate data may be fragmentary or unavailable. Causes of uncertainty include insufficient or contradictory evidence as well as human behavior. The human dimensions of uncertainty, especially coordination and strategic behavior issues, constitute a major part of the uncertainties related to climate change mitigation.

Decision-support analysis assists decision-makers, especially when no there is no optimum policy. Having a number of analytical approaches available—each with specific strengths and weaknesses—helps to keep the information content (of the climate change problem) within the cognitive limits of most decision-makers. The diversification of analytical techniques supports a more informed and effective dialogue among the involved parties. However, there are significant problems in identifying, measuring and quantifying the important variables to any decision-support analysis framework. For example, the impacts on natural systems and human health are quite difficult to measure. These variables do not have a market value that is readily quantifiable; they are often simplifications of the reality. When many decision-makers with different value systems are involved in a decision, it is helpful to identify the value judgments underpinning any analytic outcomes. This process is especially difficult when analysis aims at choices associated with high levels of uncertainty and risk.

Integrated assessments can inform decision makers of the relationship between geophysical climate change, climate impact predictions, adaptation potentials, and the costs of emissions reduction, as well as the benefits of avoided climate change damage. These assessments have methodological frameworks to deal with incomplete or imprecise data. In absence of precise data associated with climate change and GHG emissions reduction activity, the decision makers must deal with different possible scenarios, potential outcomes, and the range of possibilities for implementation protocol. Decision-makers should also establish a schedule for meeting the established targets.

DEVELOPING INTERDISCIPLINARY TEAMS

According to leadership expert, Stephen Covey: "If you get the spirit of teamwork, you start to build a powerful bond, an emotional bank account, and people subordinate their immediate wants for long-term relationships."[11] Organizations are rapidly transforming themselves into networked, cellular structures (Avery 2005). This transformation is intended to promote the exchange of knowledge, facilitate speedy responses, and generate greater adaptability in the event of market shifts. Innovative team-based systems and structures are needed to accommodate the new realities of organizations and their markets. Sharing leadership responsibilities and collaboration on projects appears to be the norm in these environments. The increasing use of teams in organizations has changed many traditional leadership roles and, consequently, empowered employees to make more autonomous decisions.

Teams are designed to complement individual skills and limitations. Self-managing teams become responsible for continuously assessing and improving their own products, services, work designs, and processes. Leadership is often elected or rotated rather than appointed. However, strong pressure falls on each individual to accept responsibility for his or her decisions and actions. One way to highlight differences in team structure and leadership is by distinguishing the extent of organizational authority allocated to each unit:[12]

- Manager-led unit—the manager directs the unit in specific tasks.
- Self-managing unit—the group manages its own performance as well as performing the task.
- Self-designing unit—members may modify the design of the unit and aspects of the context in which it operates.
- Self-governing unit—has all the above responsibilities and also decides what task must be completed.[13]

Notably, this list of work units increases, respectively, in the extent of decision-making and empowerment. The highest level is found in self-governing units. Self-governing units are coordinated through extensive communication networks and require a highly skilled workforce. Self-governing units become involved in organizational strategy, aligning their tasks and responsibilities with the organization's strategic direction. Such units are accountable to other teams they interact with; they are commonly appraised on their quality and timeliness. Appraisals reemphasize the importance of satisfying both internal and external customers.

While silos and boundaries are continually present in team environments, ideas about them change. Organizations try to develop lateral coordination among teams. In doing so, coordination occurs less frequently from managers and more frequently among groups. Thus, new leadership styles surface that embrace the differences within and among groups. In many team environments the leadership role is similar to that of a facilitator or coach. Team leaders are also caretakers. They help their teams achieve goals by providing them with instructions, encouragement, and resources. A leader's central role allows them to effectively support the team, helping members develop the necessary skills and resolve conflicts.

One innovative approach to organizational change, with an emphasis on preparing and adapting to climate change, is proposed in "Creating a Climate for Change."[14]

The "wheel of change" toward climate and environmental sustainability in organizations begins in different places.[15] Such organizations sequence interventions in different ways. For some organizations, a successful intervention hinges on a major crisis; others require strong, charismatic leadership. Still, all tend to create a sustainable imperative. Clearly defined missions and goals, along with guidelines for how to achieve them, are needed to implement green initiatives. Those missions and goals should present a compelling vision. Then, a sustained effort is required to ensure that they are pursued. For instance, company-wide use of renewable energy should be followed by supportive decision-making about specific purchases, processes, and practices.

Rearranging organizational systems requires broad involvement of organizational members and key external stakeholders. A variety of short-term sustainability "transition teams" could be organized to support green initiatives. The dynamics of organizing and operating creative teams is a complex issue. Each team must be clear about what it is striving to achieve, the role of each member, and the rules that will guide its operations. Furthermore, teams evolve as members analyze, plan, and implement strategies that support environmental sustainability. Team responsibility requires a collective mindset, supported by appropriate organization culture.

SUMMARY

In all measures of dealing with global climate change, whether through the implementation of government policy or engagement in advanced technological research, there is a notable reliance on human resources. In order to effectively mitigate carbon emissions, humans must integrate efforts and support leaders who are willing to progressively address global climate change. In many cases, the ability of a committed, environmentally conscious human agent to make a difference depends on his or her respective medium and its organizational culture. Limited models exist to guide such organizational changes. Even fewer apply to the type of interdisciplinary, international collaboration that is required to thwart global climate change. Finally, a high level of cooperation is particularly difficult to achieve without a supranational body overseeing the regulation of carbon emission reductions. Enormous pressure exists for proponents of mitigation strategies to illustrate and prove sufficient benefits. Proponents must secure voluntary participation and create a common vision of responsible behavior. It can be a daunting task.

This chapter demonstrates methodologies available for teams, organizations, and governments to make incremental changes and eventually incorporate the principles of sustainable development. These methodologies can be used to redirect governments and organizations toward the goal of reducing carbon emissions. The ability of such entities to change and procure change is partially attributable to recent trends in organizational culture and the propagation of globalization. Organizations are rapidly transforming themselves into networked, cellular structures; team-based work is becoming the norm. This transformation is intended to facilitate speedy responses, enable exchange of knowledge, and generate greater adaptability. Moreover, it empowers employees and teams in a pursuit of more environment-friendly solutions. Improvements in communication and technology, as well as the expansion of markets, foster the exchange of information. These technological advancements make organizations and individuals effective advocates for more socially responsible, green practices.

This chapter identifies the aspects of risk and uncertainty. Both are closely associated with the externalities of global climate change and seen as potential obstacles in the development of a related mitigation approach. Indeed, establishing clear missions and goals for an organization, along with guidelines for how to achieve them, is needed for green change. In the absence of precise data and projections, decision makers must deal with a range of possible scenarios, outcomes, and possibilities for implementation protocol. *How do leaders make responsible, comprehensive decisions when opponents stress uncertainties of the situation? They do this by stressing the importance of their vision and by focusing on the issues associated with risk mitigation and by accepting incremental changes in policies and progress.* Organizational change is often difficult to achieve; admittedly, change management is a complex process. This is especially true when organizations attempt to alter policies and practices regarding climate change because the potential economic benefits are pitted against immediate bottom-line pressures and costs. Therefore, climate advocates and climate scientists must offer visionary, progressive messages; finding the best possible organizational leaders and allies rather than waiting for ideal allies. They must be flexible and use creative approaches that provoke and guide organizational change.

Endnotes

1. Cashman, K. and Forem, J. (2003). *Awakening the leader within: A story of transformation*. Wiley & Sons, Inc.
2. Avery, G. (2005). *Leadership for sustainable futures*. Edward Elgar Publishing Limited.
3. United Nations Development Program (2005).
4. IAEA (2008, December 17).
5. U.S. Environmental Protection Agency.
6. California Public Utilities Commission (2008, April 10). http://docs.cpuc.ca.gov/Published/News_release/81168.htm, accessed 5 January 2009.
7. California Institute for Climate Solutions Executive Summary, Appendix A. http://docs.cpuc.ca.gov/word_pdf/FINAL_DECISION/73232.pdf, accessed 22 December 2009.
8. State of California's Legislative Assembly Bill 32
9. Governor of the State of California. (2005, June 1). Executive Order S-3-05. Signed by Arnold Schwarzenegger. http://gov.ca.gov/executive-order/1861/.
10. Jhaveri, A. (2006). *Effective leadership for sustainable development in the public sector.* Doctoral Dissertation.
11. Covey, S. (2007). *Principle centered leadership*. Franklin Covey Company. p. 46.
12. Hackman, R. (2004). *Leading Teams.*
13. *Ibid.*
14. Moser, S. and Dilling, L. (2007). *Creating a climate for change*. Cambridge University Press.
15. Doppelt, B. (2003). *Leading change toward sustainability: A change-management guide for business, government, and civil society*. Greenleaf Publsishing Limited.

Chapter 12

Action Plan for Implementing Carbon Reduction

"One thing is clear: the fate of cities will determine more and more not only the fate of nations but also of our planet. We can afford to ignore the issue of the sustainable management of our cities only at our own peril."

Elizabeth Dowdeswell (1996),
*former United Nations Under-Secretary General and
Executive Director, United Nations Environment Program*[1]

DEVELOPING A CARBON REDUCTION MANAGEMENT PLAN

To understand a carbon reduction management plan it is critical to examine the global environmental movement that has impacted humans during the past 35 years or so. The principles set out in the non-binding 1972 Stockholm Declaration on the Human Environment were developed in numerous subsequent formal and informal agreements. They were identified at the Earth Summit in Rio de Janeiro in 1992 where world leaders signed conventions on climate change, bio-diversity, and desertification. They also adopted Agenda 21, a wide-ranging blueprint for action to achieve worldwide sustainable development. The Earth Summit concept of "think globally, act locally" inspired action from governments, community groups, and individuals around the world. The Kyoto Protocol on Climate Change was negotiated in late 1997 in Japan. Subsequently, the Earth Summit was followed by the World Summit on Sustainable Development in Johannesburg, South Africa, in 2002. There, governments agreed on a non-binding Plan of Implementation that was supported by the launch of a number of multi-stakeholder partnerships to implement specific actions. The United Nations Commission for Sustainable Development later reviewed the Johannesburg Commitments on Sustainable Energy.

In the case of climate change, it is clear that there are dimensions and overlapping approaches to international cooperation (Stern Review Report 2006). Transparency and a shared understanding of actions is required across all of these dimensions, including emissions reductions, the scope and level of carbon prices and policies, investments in innovation, parallel and coordinated approaches to standards and regulations, commitments to international cooperation on the deployment and diffusion of relevant technology, as well as international support for adaptation. Global public concern and awareness about climate change are growing rapidly. They influence and sustain international cooperation, national aspirations, and private sector leadership on climate change.

Therefore, the development of a carbon reduction management plan must incorporate much of the structural content, conceptual framework and guiding principles regarding carbon reduction, irrespective of the scope or type of the plan. One example of a comprehensive action plan was developed and implemented by the City of Austin, Texas. It provides an excellent, replicable model for similar urban settings around the world.

The Austin Climate Protection Plan (ACPP) is perhaps the most comprehensive GHG emissions reduction plan undertaken by a city.[2] The Austin City Council, Austin Energy (a locally-owned electric utility), and citizen stakeholders, are leaders in energy efficiency, renewable energy, and innovative technologies. The ACPP vision for the future requires shifting from the traditional "linear energy" flow to a "multiple-fuel, multi-directional" strategy.[3] Some of the major targets include:

- Making municipal facilities, fleets, and operations carbon neutral by 2020.

- Having city facilities be 100% renewable by 2012.

- Ensuring the optimum level (maximum accomplishment based on efficiency and cost-effectiveness) of city department reductions in GHG emissions and energy consumption.

- Educating, motivating, and supporting Austin's 10,000 city employees to reduce their personal carbon footprints through outreach, training, and incentives.

Actions related to these targets as identified in the ACPP are being implemented. The ACPP includes a green building program as part of its plan. With a goal of providing 30% of its energy from renewables by 2020, the city has already achieved a 5% target.[4] The ACPP also includes a green power program, known as GreenChoice, with a current subscription of 665 million kWh annually.

An effective methodology for developing a climate protection plan should have the following five specific components within its framework to address management of carbon emissions:[5]

1) Conduct a greenhouse gas emissions inventory and projection

2) Establish an emissions reduction target

3) Develop a local action plan to reduce emissions

4) Implement the local action plan

5) Monitor progress and report on results

REAL WORLD RESULTS OF CARBON REDUCTION PROGRAMS

Before providing specific examples of successful carbon reduction programs and innovative approaches that result in significant measurable results, it is important to articulate the task that lies ahead to mitigate, adapt, and reduce the ever-increasing threats of global climate change. In order to prevent a worst-case climate change scenario (defined as limiting CO_2-equivalent atmospheric concentration to 500-550 ppm by 2050), teamwork will be required at the local, regional, national, and international levels. The effort to not exceed the 500-550 ppm-level needs to involve public and government entities, private industry, non-governmental organizations (NGOs), and community partnerships. At the same time, realistic strategic plans for CO_2 and GHG emission reductions will need to be phased into short-term (2010), mid-term (2025), and long-term (2050) time frames. These timeframes will facilitate realistic mitigation strategies to be implemented and allow effective monitoring of the progress toward near-term, mid-term and long range goals.

This teamwork will require the active participation of community leaders, decision makers, government agencies, business and industry representatives, and financial institutions. To ensure broad involvement in this complex undertaking, it is recommended that the principles of sustainable development (i.e., environmental stewardship, economic prosperity, and social justice) become the foundation for carbon reduction policies, strategies, and technologies. These principles serve as the conceptual framework for the shared vision and societal benefits that can be advanced by a focus on sustainability. Such visionary approaches have been responsible for the successes of other similar programs, projects, and partnerships.

The United Nations Framework Convention on Climate Change and the Kyoto Protocol embody the core principles of a multilateral response. The Kyoto Protocol (entered into force in February 2005) set out an approach for binding international action and agreed specific commitments up to 2012. As a result, climate change is becoming central to international economic relations, along with issues such as trade, development, and energy security. A range of institutions and arrangements support coordinated or parallel action on energy policy and land-use change.

The majority of the world's largest economies now have goals in place to reduce carbon emissions, to decrease their energy intensity, to increase their use of renewable energy, and to decrease deforestation. Countries have adopted a range of goals, hoping to result in significant emissions reduction upon implementation. Ten of the world's largest economies have adopted aggressive climate change and clean energy goals—Brazil, China, France, Germany, India, Italy, Russian Federation, UK, and the U.S.[6] Many countries, regions, and cities have adopted carbon-neutral approaches that complement and go beyond action under the multilateral framework.

International companies and multi-national corporations are taking a lead in demonstrating how profits can be increased while reducing emissions from industrial activity. Visions for a zero carbon society vis-à-vis private sector leadership include major automotive, power, energy, and financial industries like Toyota, Avis Europe, Vattenfall, Alcan, and HSBC. Each of these corporations is attempting to accomplish goals for CO_2 and GHG emissions reductions from their activities.

Many cities, states, and the regions within the U.S. have not only adopted climate change and CO_2 and GHG emissions reduction action plans with specific targets, but also have become leaders in helping others to establish similar programs. According to the U.S. Conference of Mayors' Climate Protection Agreement, led by Mayor Greg Nickels of Seattle, more than six hundred mayors of small, medium, and large cities in the U.S. have committed to reducing their respective urban CO_2 and GHG emissions. This commitment represents constituencies of over 90 million Americans. The Climate Protection Agreement is particularly important since cities account for nearly 78% of all GHG emissions.

The City of Portland, Oregon, and the West Coast Governments' partnership involving the states of Washington, Oregon, California, Arizona, New Mexico, among others, provide prime examples of collective action. In 2001, Portland set an overall goal of reducing CO_2 emissions 10% below 1990 levels by the year 2010. An evaluation of Portland's efforts between 1990 and 2004 showed impressive results. CO_2 emissions were reduced by 12% per capita—possibly the largest reduction anywhere in the United States.[7] Similarly, many residents, businesses, and governments in the states of Washington, Oregon, and California are "thinking globally and acting locally" regarding climate change.[8] The topics of climate and energy bring many stakeholders together. A focus on impacts includes interested parties from agriculture and forestry, transportation and cement production, water and waste management, as well as public health, to the mitigation effort. As West Coast states take the lead on developing regulated emission reduction strategies in their region, their efforts hold potential to facilitate developments in the global arena.

REPLICABLE MODELS

U.S. EPA's Energy Star program addresses the environmental issue of global climate change through increased energy efficiency. The primary goal is to develop voluntary partnerships, reduce the energy intensity of manufactured products and buildings, subsequently lowering GHG emissions through reductions in energy use. An objective of this program is to achieve a target set in 2002, which is to reduce GHG emissions 18% by 2012. Annual goals are revisited every four years, as part of the Climate Action Report (CAR) process. Ideally, this reporting process is consistent with the reporting requirements of the United Nations Framework Convention on Climate Change. The programs under this initiative are scored using the Program Assessment Rating Tool (PART). Future goals (through 2020) are formalized in CAR policies and projections. The U.S. EPA also provides support, training, and tools to aid in verifying performance and results.

Business planning processes provide a means of defining performance indicators that contribute to carbon reduction goals. Examples of performance indicators include:

- Number of partners
- Number of product specifications added/or revised
- Percent of public awareness
- Number of products sold by product type (compared to prior periods)
- Number of new homes built
- Number of buildings benchmarked
- Energy efficiency improvements in these buildings
- Number of industry performance indicators released

From 2000 to 2004, substantial increases in utility savings and emissions equivalents throughout the entire U.S. can be attributed to energy efficiency improvements. Such results confirm the following achievements for every U.S. dollar invested in the Energy Star program:[9]

- Reducing GHG emissions by 1.0 MTCE or 3.7 tons of CO_2;
- Saving businesses and consumers more than $75 in energy on their utility bills;
- Creating more than $15 in private sector investment; and
- Adding over $60 to the U.S. economy.

With an eye on environmental best practices and transportation and logistics sustainability, Deutsche Post Work Net (DPWN), the parent company of DHL in Bonn, Germany, has an initiative focused on reducing its carbon footprint by 30% by 2020. This initiative, called the "GoGreen Program," focuses on reducing the company's carbon footprint for every letter mailed, container shipped, and meter of warehouse space used.

As a logistics company, with more than 500,000 employees in 220 countries, DPWN feels strongly about its corporate social responsibility in terms of societal benefits and environmental stewardship. DPWN's steps to reduce its carbon footprint include:[10]

- Replacing 90% of its air fleet with modern, more fuel-efficient aircraft including 48 Boeing 757 special modern freighter cargo planes, which require 20% less fuel per ton transported than the Boeing 727 predecessors;

- Adding aerodynamic winglets to six cargo aircraft to increase the amount of lift generated at the wingtip, which raises aircraft efficiency by 2 to 5%;

- Increasing the use of alternative fuels and propulsion units with biogas-fueled heavy-duty trucks, delivery vans;

- Using vehicles with hybrid engines and making changes in transportation methods (e.g., air, sea, and rail); and

- Optimizing routes using intelligent traffic guidance systems and other route-planning technologies to reduce fleet fuel consumption.

The company also cited environmental technologies it will implement to improve the energy efficiency of sorting centers and warehouses.[11] One example is increasing the proportion of regenerative energy DPWN uses in its global operations at the DHL cargo hub located at the Leipzig-Halle Airport—a change that provides an annual savings of 4,000 metric tons of carbon.[12] It does this by using electricity produced by solar photovoltaic cells, CHP, and power generators for internal energy requirements.

With regard to other replicable models for carbon reduction, significant discussion has been provided under "Modeling Systems" in Chapter 10 of this book. However, it is worthwhile to summarize the key aspects of this important concept for CO_2 and GHG emissions reduction, specifically regarding "bottom-up" and "top-down" concepts.[13]

The bottom-up approach is based on assessment of mitigation options, emphasizing specific technologies and regulations. These typically involve studies of economic sectors that consider the macro-economy as static. Sector estimates are aggregated to provide an estimate of global mitigation potential. In particular, bottom-up studies are useful for the assessment of specific policy options at the sector level (e.g., options for improving energy efficiency).

The top-down concept assesses the economy-wide potential of mitigation options. This analysis uses globally consistent frameworks and aggregated information about these options and captures macro-economic and market feedbacks. The top-down studies are useful for assessing cross-sector climate change policies, such as carbon taxes and stabilization.

Bottom-up and top-down models have become more similar since the Third Assessment Report (TAR) of the IPCC. The TAR incorporated more technological options (top-down) and more macro-economic and market feedbacks (bottom-up), and adopted barrier analysis into the various modal structures.[14] Other features that distinguish between these two models are:

- The electricity savings that are allocated to the power sector in the top-down models compared to end-use sectors;

- The top-down models indicate a higher emission reduction such as extraction and distribution, reductions of other non-CO_2 emissions, and reductions through the increased use of CHP;

- In bottom-up estimates, fuel substitution is assumed only after end-use savings, while the top-down models adopt a more continuous approach;

- In the building sector, the estimates of reduction potentials are lower for top-down vis-à-vis bottom-up assessments;

- In case of agriculture and forestry, reduction estimates are higher from bottom-up vis-à-vis top-down models; and

- The top-down models generate higher reduction potentials in industry than from the bottom-up assessment.

The actual programs and models described earlier in this chapter, along with their results, rely on measurement methodologies and proto-cols. Measurement protocols quantify and periodically report the carbon or CO_2 and GHG emissions reductions targets, consistent with the site-specific registry requirements. This process is not only helpful to specific companies, corporations, agencies and organizations, but also offers the potential to monitor the changes in emissions in a transparent manner. However, this new field of emissions accounting has not been bound by any regulated compliance arrangement by organizations such as the Financial Accounting Standards Board or the International Accounting Standards Board. Measurement and verification protocols being used today for CO_2 accounting have been established by the various regional regulatory entities.

WHAT TO DO WHEN STRATEGIES FAIL?

When carbon reduction strategies face dilemmas associated with failure to meet established reduction goals, the answer is to take proactive and feasible measures to avoid such failures in the future. According to the 2007 report "Carbon Reduction Strategies: Choosing the Right Path for your Business," strategies tend to become most effective when they are implemented as part of an overall carbon reduction or climate change plan.[15]

Carbon reduction strategies contain specific actions that can be taken to reduce an entity's carbon footprint. These apply equally to public and non-profit sectors. These strategies fall into three basic categories:

1) **Reduce**—the first and most important step in reducing the footprint is through energy conservation and efficiency measures, such as turning off lights and equipment when not in use and minimizing auto and airplane travel. This category also encompasses the use of alternative fuels like biodiesel and ethanol.

2) **Switch**—Another step in carbon reduction is switching from fossil fuel-based energy sources like coal, natural gas, and oil to those utilizing renewable energy sources such as solar and wind power.

3) **Offset**—Offsets are often used to help organizations meet CO_2 and GHG emissions reduction targets when the cost of internal reductions is excessive. While an offset does not directly reduce the purchaser's emissions, it indirectly mitigates emissions by funding projects that generate renewable energy, increase

energy efficiency, capture GHG from industrial, agricultural, or landfill operations, or increase CO_2 and GHG sequestration, typically through forestry and land management. Offsets are almost always necessary if an entity is to become carbon neutral.

Therefore, the factors to be considered in selecting the right strategies are:

- The specific goals (e.g., to become carbon neutral or to reduce footprint);
- The biggest impacts (e.g., energy efficiency vs. transportation);
- Where immediate results can be achieved (e.g., small vs. large changes); and
- Other factors (e.g., target groups and implementation protocols).

SUMMARY

It is clear from this chapter that to be successful in carbon reduction at the local, regional, national, and international levels, each entity responsible for their implementation must develop a realistic strategic or management action plan. These plans must be comprehensive, sustainable, and inter-disciplinary in nature. There are many examples and case studies that have been identified herein, which could be used as replicable models by those planning to get involved in carbon reduction programs. It is remarkable that many cities and urban communities around the world have developed action plans for carbon reduction implementation, which are more successful than their national and international counterparts. These convincing grass-roots efforts have been briefly discussed and documented in this chapter.

Endnotes

1. Dowdeswell, E. (1996). United Nations Environment Program.
2. Muraya, N. (2008, November). *Austin Climate Protection Plan*. Austin Energy.
3. Muraya, N. (2008, March). Austin climate protection plan "possibly the most aggressive city greenhouse-gas reduction plan." *Energy Engineering*, 105 (2).
4. *Ibid.*
5. Intergovernmental Panel on Climate Change (2007, November). Report on *Mitigation of climate change*. http://www.ipcc.ch/ipccreports/ar4-wg3.htm
6. Stern Review (2006). *International collective action*.
7. Global Warming Progress Report (June, 2005). www.portlandonline.com, accessed 28 September 2006.

8. Dilling, L. and Moser, S. (2007). *Creating climate for change*. Cambridge University Press.
9. U.S. Environmental Protection Agency (2005, November). *Measuring Energy Star program results*.
10. Deutsche Post World Net (2008). Accepting our social responsibilities. http:// www.dpwn.de/dpwn?skin=hi&check=yes&lang=de_EN&xmlFile=300000032, accessed 29 May 2008.
11. Deutsche Post World Net (2008, April 8). Deutsche Post World net starts global climate protection program GoGreen. http://www.dpwn.de/dpwn?tab=1&skin= hi&check=yes&lang=de_EN&xmlFile=2009784, accessed 29 May 2008.
12. Berman, J. (2008, April 16). *Breen logistics: DPWN sets goal of 30 percent carbon footprint reduction through 'GoGreen' initiative.* http://www.scmr.com/article/ CA6551866.html, accessed 20 January 2009.
13. Intergovernmental Panel on Climate Change (2007, November). Report on *Mitigation of climate change*. http://www.ipcc.ch/ipccreports/ar4-wg3.htm
14. Intergovernmental Panel on Climate Change (2001, May) *Third assessment report*
15. Cascadia Consulting Group, Inc. (2007, October). *Carbon reduction strategies: choosing the right path for your business*.

Chapter 13

Creative Financial Approaches To Carbon Reduction

"Offsets should be viewed as one of several types of financial incentives that target emissions not covered under a mandatory carbon cap. Other financing mechanisms, such as tax credits, rebates, and grants, may be more appropriate ways to encourage reductions in some types of un-capped emissions sources. That's why climate change legislation should include both a mandatory system for reducing emissions—via the "cap" part of cap-and-trade—and a suite of incentives—including offsets—that seek to reduce emissions from sources that are not restricted under the cap."

David J. Hayes (2008)[1]

There are a number of approaches to financing carbon reduction projects. Many are similar to the ways that energy efficiency and alternative energy projects have been financed in the past. However, there are difficulties in financing such projects. The problem is not a lack of funding availability but often a lack of commercially available financing.[2] Traditional financing approaches often fail to properly credit the annuity value of avoided future costs when future expenses, such as the costs of energy, can be widely volatile. Carbon taxation is used in some countries to mitigate such volatility.

Additional hurdles include difficulties in getting access to available funds from local financial institutions. This can be caused by disconnects in lending practices as they apply to energy efficiency and carbon reduction improvement projects.[3] These may include structured restrictions on capital, collateral requirements, ownership provisions and market conditions. This became increasingly evident in 2009, when lending was disrupted by a severe recession which impacted financial markets. Regardless, there are ways to overcome these hurdles by using creative financial approaches. These include programs such as the use of carbon offsets and credits, funding demonstration projects, the use of performance contracting, and creative rule structures for measurement and verification of savings.

CARBON OFFSETS AS A FINANCIAL INCENTIVE

The report "The Economics of Climate Change" explains how "cap-and-trade" systems control the overall quantity of GHG and CO_2 emissions by establishing binding emissions commitments.[4] These systems regulate contaminates from generation sources by establishing limits on the allowable levels of emissions and by requiring reductions in contaminate levels over a period of time.

Using cap-and-trade systems, an emission maximum is established for the pool of participants and each participant in the pool. Within this quantity ceiling, entities covered by the plan—such as individuals, organizations, or countries—are then free to choose how best, and where, to deliver emission reductions. The largest example of a regulated cap-and-trade scheme for GHG and CO_2 emissions is the EU's Emissions Trading Scheme. There are other national and regional emissions trading schemes, including the U.S. Regional Greenhouse Gas Initiative and the Chicago Climate Exchange (CCX). The CCX allows participants (e.g., companies who have accepted voluntary commitments to reduce emissions) to purchase carbon financial instruments (CFIs). Projects eligible for CFIs include reforestation, afforestation, and soil "carbon offsets" in the agricultural and biomass sectors. Offsets may be created through the use of conservation tillage and grass planting. Eligible projects must be enrolled through an intermediary registered with the CCX, which serves as an administrative and trading representative on behalf of the multiple individual participants. The intermediary body is known as an offset aggregator. The first sale of verified CO_2 offsets generated from agricultural soil sequestration in the U.S. using an exchange occurred in April 2005.[5] By June 2006, approximately 140,640 hectare (350,000 acres) of conservation tillage and grass plantings had been enrolled in Kansas, Nebraska, Iowa, and Missouri.[6] The measures to enhance natural soil fertility and carbon sequestration potential can also have additional benefits such as the decreased use of man-made fertilizers, reduced deforestation, improved water quality, reduced power consumption, and diminished fuel requirements for tillage.[7]

Another example of the use of carbon offsets is exemplified by the creation in 2003 of a regional cap-and-trade system, known as the Regional Greenhouse Gas Initiative (RGGI), an initiative developed by states in the Northeast and Mid-Atlantic regions (discussed in Chapter 4). The basic objective of the RGGI is to reduce GHG emissions from the electrical power sector while maintaining economic competitiveness and efficiency across the regional power market. The framework of the RGGI follows the model of earlier successful cap-and-trade systems, such as those used to reduce acid rain-producing sulfur dioxide (SO_2) emissions. Cap-and-trade programs include the following basic components:

- A decision about who is being regulated,
- Mandated reduction levels,
- Provision for the distribution or allocation of permits or "allowances" that power generators will need, and
- The structure of the trading mechanism.

The RGGI agreement regulates medium and large power plants, with nearly 75% of the allowances in each state to be distributed among the generators, while the balance is set aside for various public benefit uses. The agreement also allows the generators who reduce GHG and CO_2 emissions further than required to bank those allowances for the future or sell them to other generators. Moreover, generators who have not met their reduction goals must purchase allowances on the market or purchase a limited amount of reductions through offsets. Carbon offsets are certified financial instruments that represent emission reductions that are typically achieved outside of the power production sector. These offsets are financial tools that provide incentives for emitters of CO_2 to reduce their emissions. Offsets are typically derived from emission reductions through mitigation efforts such as energy conservation or alternative energy projects, reforestation or landfill methane recapture. RGGI is expected to cap CO_2 emissions from power plants at 2005 levels, between 2009 and 2014. From 2015 onward, the cap will decrease, so that by 2019, the states collectively will have reduced their emissions by 10% below 2005 levels.[8]

COSTS OF CARBON CAPTURE SYSTEMS (CCS)

The IPCC's 2007 Mitigation Report defined the CCS process as capturing a stream of highly-concentrated CO_2 and either storing it geologically—in the ocean or in mineral carbonates—or using it for industrial purposes. Highly concentrated streams of CO_2 are produced by coal-fired electrical generating plants. The IPCC Special Report on CCS offers a description of this technology and describes its potential applications in industrial settings.[9] Industry only rarely considers using CCS as a mitigation option because of its technical difficulties and high costs. However, assuming that the research and development (R&D) currently underway that focuses on lowering CCS costs proves successful, the application of CCS technology to industrial CO_2 sources might begin before 2030 and become widespread.

CCS is a process that is yet to be deployed at full commercial scale in the power sector, so it remains at the demonstration stage of the innovation process.[10] The IPCC Special Report on CCS suggested it could provide between 15% and 55% of the cumulative mitigation effort through 2100. Failure to develop CCS would result in a narrower portfolio of low carbon technologies and this could increase abatement costs. Recent International Energy Agency (IEA) modeling shows that, without CCS, less abatement occurs at a higher cost. Indeed, marginal abatement costs would increase by roughly 50%. Modeling work undertaken for the Global Energy Technology Strategy Program showed that removing the option of CCS more than triples the cost of stabilizing CO_2 for all concentration levels analyzed.[11] This prominent role of CCS in future mitigation efforts can be linked to the expected global growth of coal use.

Electric utilities hoping to develop CCS projects have significant financial hurdles to overcome. A single CCS demonstration project costs several hundred million dollars more than the cost of a standard power station. The IEA recommended that 10 to 15 such projects should be in place by 2015 at an estimated extra cost of $2.5 to $7.5 billion, in order to demonstrate the commercial viability of the technology.[12] This is substantially more than the $100 million annually that is currently spent on research and development for CCS technologies. While one approach might be for CCS demonstration projects to be developed by a group of countries using pooled resources, there are currently no arrangements for coordinating such efforts. There have been several announcements from governments and the private sector on planned CCS projects. The CCS projects include:

1) The U.S. Futuregen project, which is linked to the demonstration

of integrated gasification combined cycle (IGCC) coal generating technology and has lost federal funding support.

2) British Petroleum's proposed project at Peterhead, which includes a 350 MW hydrogen plant, capturing 1.2 million tons of carbon each year; and a feasibility study by RWE for a post-combustion techniques in a 1,000 MW coal plant in Tilbury, UK.

3) A Japanese proposal to capture a sixth of all their emissions by 2020.

4) Vattenfall's 30 MW pilot coal plant in Germany which began operation in 2008.[13]

5) A geologic storage pilot project in the Otway Basin in Western Victoria planned by a public-private research organization in Australia. A liquefied natural gas (LNG) project at Gorgon (North West Shelf), and the Stanwell ZeroGen IGCC-CCS project, are at the proposal stage.

6) The EU has an initiative seeking to develop a CCS plant in China.

Building on these announcements, enhanced coordination of national efforts could allow governments to allocate more support to the demonstration of a range of different projects. This could show different pre- and post-combustion carbon capture techniques by different generation plants, since the appropriate technology may vary according to local circumstances and fuel prices. Another approach is to focus on the best ways to make new plants "carbon capture ready," by building them so that retrofitting CCS equipment is possible at a later date.

FINANCIAL SOLUTIONS TO CARBON REDUCTION

Diverse evidence indicates that carbon prices in the range $20 to $50 per ton of CO_2 ($75 to $185 per ton of carbon) emitted, reached globally by 2020-2030 and sustained or increased thereafter, would provide adequate economic incentives to deliver deep emission reductions by mid-century. This trend will be consistent with stabilization at 550 ppm

CO_2-equivalent, if implemented in predictable fashion.[14] Such prices would deliver these emission savings by creating financial incentives large enough to switch ongoing investment in the world's electricity systems to low-carbon options, to promote additional energy efficiency, and to halt deforestation and reward afforestation (which is the process of establishing new forests). The emission reductions will be greater to the extent that carbon prices are accompanied by expanding investment in technology research, development, and demonstration (RD&D) and targeted market-building incentives.

Acting now to ensure that the current wave of investment in fast-growing economies which incorporate energy efficiency and low-carbon technology, will ultimately reduce the global cost of stabilizing GHG in the atmosphere.[15] Private sector resources for energy sector investment far outweigh those funds available from governments and multi-lateral institutions. Middle-income countries, where the bulk of future GHG emissions growth is concentrated, have access to capital from the private sector. Regardless, public sector resources are an important lever to channel flows of domestic and international private sector investment into carbon reduction activities.

Successful and effective financial solutions to energy and carbon reduction programs include the Global Environmental Facility (GEF) and the CDM. The most profitable projects to generate carbon credits under Kyoto's Clean Development Mechanism (CDM) are being exhausted after a two-year flurry by Western financial houses to secure them.[16] While the most lucrative deals reflect activity to eliminate emissions of many potent greenhouse gases, there is concern that they limit financing for projects which provide more long-term solutions to fight climate change, particularly the development of clean energy.[17] The GEF has a strong history of financing energy efficiency and renewable energy programs. The collective impact of these programs is small relative to the challenge of reducing CO_2 and GHG emissions. Regardless, the CDM provides an important channel for private sector participation in financing low-carbon investments in developing nations.

Since its inception in 1991, the GEF has provided $6.2 billion in grants and generated over $20 billion in co-financing from other sources to support over 1,800 projects that produce environmental benefits in 140 developing countries.[18] The World Bank has recently suggested that the GEF could play an enhanced role in encouraging technology transfer and lowering the cost of the low-carbon technologies that are relevant to the priorities of developing countries. To enhance its role, the GEF would require a two to three-fold increase in current financing in order to ensure sustained market penetration of energy efficiency and renewable energy technologies over the next decade. Financing a strategic global program to support the reduction in costs of pre-commercial, low GHG emitting technologies such as CCS, solar thermal, or fuel cells, would require more than a ten-fold increase.

Carbon offset credits are tradable credits earned for investing in projects to reduce greenhouse gas emissions. In CDM terms, these credits are called certified emissions reductions (CERs) and are typically generated in developing countries. One CER represents an emission reduction of one metric ton of CO_2. Such credits are valuable to governments and companies in developed nations because the Kyoto Protocol allows them to use these credits to offset their own greenhouse emissions and help them meet a portion of their domestic reduction targets.[19] Buying credits, or paying for them to be generated, can be less expensive than reducing emissions at home. Traded amounts under the CEM were equivalent to 717 metric tons in 2005, increasing to 1.0 billion by 2006.[20] Over 1.5 billion annually are expected to be generated by the end of 2012, when the targets under Kyoto I expire.[21]

The CDM has been criticized for rules which create opportunities for investments in low cost projects at the expense of more costly (and more lucrative) projects to create renewable energy and reduce carbon dioxide emissions.[22] A meeting of Asian and European nations in 2006 identified the concentration of CDM projects in countries such as China and India. Many want to see more projects in renewable energy in poor African countries to spur sustainable economic development.[23]

Again in 2006, Europe's Climate Change Capital, Deutsche Bank and Centrica were claiming to have sealed the largest CDM transaction arranged purely by private companies. It was for an HFC project in China at the Zhejiang Juhua chemical plant to provide 29.5 million tonnes of CO_2-equivalent reductions annually.[24] At the prevailing prices in 2006, this is worth 300 million Euros (the equivalent of roughly $500 million) a year or more to investors.[25] In 2006, eight of the ten biggest CDM projects registered were for HFC destruction. Once such projects have been completed, companies are more likely to switch to renewable energy projects.

There are a number of greenhouse gases more potent than CO_2 that offer emission reduction opportunities. Among the most profitable is the destruction of nitrous oxide, which has 310 times the global warming potential of CO_2. Nitrous oxide is generated by fertilizer and nylon production. Next is avoiding the release of methane gas, 21 times as potent CO_2, from landfill, coal mine and gas-flaring projects. Investment firms in Europe and Japan are turning their attention to these.

A meeting of the UN climate change convention and the Kyoto nations took place in Nairobi in November of 2006. On the agenda was creating a successor to the Kyoto treaty that expires in 2012. While there appeared to be strong support worldwide for the CDM to continue, moves to reform the mechanism were unsuccessful. There were calls for emissions-lowering fossil fuel technology, such as carbon capture and storage (CCS) at CO_2-emitting power plants, to be eligible for carbon credits.

Ultimately, the greatest weapons in the battle against human-induced global warming will be technologies that reduce CO_2 emissions from energy generation. The sooner clean energy approaches, such as renewables and near-zero fossil fuel technologies, can be implemented, the more possible it will be to meet the challenge of reducing greenhouse gas emissions by 80 percent by 2050.

ENERGY SAVINGS PERFORMANCE CONTRACTS (ESPC)

Energy service companies (ESCOs) have been in existence for over a century. They are companies that specialize in identifying and implementing cost-effective energy efficiency upgrades to facilities. Energy savings performance contracts (ESPC) are a means of implementing and financing energy efficiency upgrades by partnering with an energy service company (ESCO) that guarantees savings that result from upgrading building systems. In an ESPC, the ESCO assumes some or all of the financial risk of the investment by guaranteeing that the host organization (e.g., a university, school district, governmental entity, industrial concern, hospital, etc.) will realize the anticipated savings.

Energy savings performance contracts are actually financial tools that are used to implement energy and water cost reduction projects. Professional energy assessments (or audits) identify opportunities to improve energy efficiency in existing facilities. These projects begin with an energy and water reduction assessment to consider the types of improvements that would reduce regularly recurring utility costs. Energy assessments provide a listing of opportunities with information on the scale and scope of the various improvements available to reduce operating costs. Using the ESPC process, facility owners can use the assessment to determine the energy savings and financial impact of their investments in efficiency and alternative energy improvements and to provide economic justification for facility upgrades. In some cases the ESCO will also assume the risk of financing the improvements themselves.

From the assessments, building owners can select a bundle of improvements to include in an ESPC project. Typically, these opportunities involve lighting system upgrades, mechanical or HVAC control replacements, building envelope improvements, hot water or process efficiency improvements, water and sewer cost reduction measures or alternative energy applications. Power production and alternative energy generation may also be included.

ESPCs rely on subsidizing projects with utility savings and other costs avoidance, using measurement and verification procedures to document savings, and the provision of a third-party guarantee of savings that covers any shortfalls in projected savings. ESPCs provide an important mechanism for many cash-strapped organizations to upgrade building systems and equipment. From the building owner's perspective, future recurring utility costs are reduced by the improved efficiencies of the newer or upgraded equipment. The estimated reductions in future utility costs are used to finance the improvements and any shortfalls in projected savings are covered by the savings guarantee. The energy services provider is paid for implementation of the project and managing the measurement and verification process.[26]

CREATING A "WIN-WIN"
ENERGY-EFFICIENCY SAVINGS PLAN

Here's how ESPC projects are generally structured:

1. An ESCO is selected.

2. An agreement is signed to proceed with the planning phase of the project.

3. An investment-grade audit is performed.

4. The owner approves the measures to be implemented.

5. An ESPC is entered into by the owner and the ESCO and financing is secured for the upgrades.

6. A turnkey project is implemented under the oversight of the ESCO.

7. Operations and maintenance procedures are established to manage the efficiency of the facilities.

8. A measurement and verification (M&V) plan is developed and procedures implemented in accordance with established energy-engineering protocols. Baseline measurements obtained during the energy assessment phase are used for savings comparisons.

9. Payments are made to the ESCO and the financier.

10. Annual reconciliations are performed to determine actual savings. If savings are not realized, the ESCO is responsible for paying the owner of the facility the difference between the guaranteed dollar savings and the actual dollar savings.[27]

Large investments in ESPC projects are occurring. By 2006, the U.S. federal government had entered into over 400 performance contracts worth about $1.9 billion ($3.5 billion including financing using private sector investment), to guarantee energy savings of $5.2 billion through reductions in utility bills. The net benefit to the government was $1.7 billion.[28] Local governments, such the cities of Baltimore, Maryland, and Bloomington, Indiana, have successfully used this process as well. In 2008, the City of Covington, Kentucky, used an ESPC program to qualify a guaranteed savings of $2.25 million over project life.[29] While their project was implemented in only six months, it required much longer to plan. According to Tom Logan, Covington's Director of Public Improvements, "it looked too good to be true. It took two years to educate ourselves and city officials on the program's technical and financial components. Officials had a difficult time understanding how a contractor could guarantee savings in the millions of dollars"[30]

ESPC investments in energy-efficiency measures can be used to implement carbon reduction improvements in facilities and processes.

ESPC projects that improve the efficiency of the use of carbon-based fuels or use substitute forms of energy (such as solar or wind energy projects) provide carbon mitigation potential. Software tools that analyze and assess the savings from these projects are capable of simultaneously calculating the carbon emission reductions from these projects. The documentation generated from the analysis can be used to calculate the value of carbon offsets available from cap-and-trade regimes in local currencies. The market value of the offsets generated also provides yet another revenue stream for project financing.

SUMMARY

There is vast potential for financial incentives to target emissions and procure the consumer-driven changes necessary for most mitigation strategies. Whether such incentives are created by the market or supported by governments, they can be remarkably effective tools in the fight to forestall and avoid global climate change. In particular, this chapter illustrates the capacity of the private sector to promote carbon reduction strategies through investment and participation in fiscal endeavors, such as new technologies, CDM mechanisms, ESPC programs and nascent carbon markets.

Such developments represent an integral part of reduction strategies; they contribute to mitigation efforts by attaching a monetary value to carbon emissions. Cap-and-trade systems create a quantity ceiling for emissions. Thus, the generators who reduce GHG and CO_2 emissions further than required can sell the remaining allowance. Likewise, carbon offsets—the promotion or maintenance of biological processes or natural features that sequester carbon—can also be translated into a monetary value. For example, the first sale of offsets generated from agricultural soil sequestration occurred in 2005.

Other sequestration techniques are in the process of being developed and revised for widespread use in the United States. CCS is a process that remains at the demonstration stage of the innovation process because of its high costs. However, international agencies (e.g., the IEA) are calling for further funding to support R&D and pilot initiatives. The great expense of CCS implies that countries with many large point source emitters (e.g., power plants and other industries) in close proximity to storage sites are the most likely candidates for cost-effective CCS programs. The financial restraints and possible risks involved in CCS procedures signify the need for government investment and regulation in developing technologies. This suggests the need for joint, concerted efforts among proponents in developed nations like the United States, where the appropriate technology, infrastructure, and geographic features make CCS a possibility.

Altogether, the potential for financial solutions to eliminate the threat of global climate change, stall the growth of GHG emissions, protect the environment, save money through energy-efficiency measures and encourage the development of new industry is evident. Evidence indicates that if carbon prices reached $75 to $185 per ton by 2020, then they would deliver substantial emission reductions by mid-century. Knowledge about the possible impact of fiscal decisions should encourage every agent to act responsibly in their investment decisions and financial pursuits. People need to become more aware of how finances impact anthropogenic carbon emissions. This chapter offers a sense of empowerment and a dose of accountability in its underlying message; it asks each person to make a conscious decision to be part of the problem—or to be the solution to it.

Energy savings performance contracts (ESPC) are a means of implementing and financing energy efficiency upgrades. ESPC projects are subsidized with utility savings and other cost avoidance. An energy service company guarantees savings that result from the facility upgrades. ESPC projects that improve the efficiency of the use of carbon-based fuels or use substitute forms of energy provide carbon mitigation potential. Software tools are available calculate the carbon emission reductions from these projects—and simultaneously assess the savings created by them.

Endnotes

1. Hayes, D. (2008, February 28). *Getting credit for going green.* http://www.american-progress.org/issues/2008/03/carbon_offsets_report.html, accessed 20 April 2008.
2. Dreessen, T. (2008, July 18). *Scaling up energy efficiency financing.* Paper presented at Wilton Park, UK.
3. *Ibid.*
4. Stern Review Report (2006). *The economics of climate change.* Cambridge University Press.
5. Intergovernmental Panel on Climate Change (2007, November). Report on *Mitigation of climate change*, chapter 8, 'Agriculture'.
6. Stern Review (2007). *The economics of climate change.* http://www.rainforestcoalition.org/documents/Chapter25Reversingemissions.pdf, accessed 19 January 2009. p. 545.

7. International Soil Tillage Research Organization (ISTRO). *Reversing emissions from land use change*, chapter 25, found in the *Stern Review* (2007).

8. Dilling, L. and Moser, S. (2007). *Creating a climate for change*, chapter 26. Cambridge University Press.

9. Metz, B. (2007). *Climate change 2007*. p. 460.

10. Stern Review (2006). *The economics of climate change, Part VI: International collective action.* http://www.hm-treasury.gov.uk/d/Chapter_24_Promoting_Effective_International_Technology_Co-operation.pdf.

11. Global Energy Technology Strategy Program (2005). http://www.pnl.gov/gtsp/research/model.stm, accessed 29 May 2008.

12. Stern Review (2006). *The economics of climate change, Part VI: International collective action.* http://www.hm-treasury.gov.uk/d/Chapter_24_Promoting_Effective_International_Technology_Co-operation.pdf, p. 526.

13. RWE and Vattenfall are EU energy companies that design and construct CO_2 capture and sequestering facilities.

14. Intergovernmental Panel on Climate Change (2007, November). Report on *Mitigation of climate change.* http://www.ipcc.ch/ipccreports/ar4-wg3.htm.

15. Stern Review (2007). *The economics of climate change.*

16. Carbon Positive (2006, September 13). *Kyoto's CDM approaches the crossroads.* http://www.carbonpositive.net/viewarticle.aspx?articleID=423, accessed 24 December 2008.

17. *Ibid.*

18. Global Environment Facility (2007). *About the GEF.* http://www.gefweb.org/interior.aspx?id=50, accessed 29 May 2008.

19. *Ibid.*

20. Graham-Harrison, E. (2006, October 17). *World carbon market leaps in 2006, China share down.* http://www.planetark.org/dailynewsstory.cfm/newsid/38696/story.htm, accessed 26 December 2006.

21. Carbon Positive (2005, September 12). *What are CERs?* http://www.carbonpositive.net/viewarticle.aspx?articleID=34, accessed 24 December 2008.

22. Carbon Positive (2006, September 13) *Kyoto's CDM approaches the crossroads.* http://www.carbonpositive.net/viewarticle.aspx?articleID=423, accessed 19 January 2009.

23. *Ibid.*

24. *Ibid.*

25. *Ibid.*

26. The International Performance Measurement and Verification Protocol (IPMVP) provides a framework for measurement and verification methodologies. Information regarding this Protocol is available at www.efficiencyvaluation.org.

27. Hansen. S. (2006). *Performance contracting: expanding horizons.* Lilburn, GA: The Fairmont Press.

28. United States Federal Energy Management Program. *Fact sheet: energy savings performance contracting,* http://www1.eere.energy.gov/femp/pdfs/espc_fact_sheet.pdf.

29. Public Works Magazine (2009, February). *Kilowatt killers.* p. 16. http://www.pwmag.com/industry-news.asp?sectionID=0&articleID=879934, accessed 25 February 2009.

30. *Ibid.,* p. 18.

Chapter 14

Implementing Successful Carbon Reduction Programs

"Is it enough for a scientist simply to publish a paper? Isn't it a responsibility of scientists, if you believe that you have found something that can affect the environment, isn't it your responsibility to actually do something about it, enough so that action actually takes place?"

Mario Molina, Nobel Laureate Chemist (2001)[1]

STRATEGIC PLANNING

How are planning strategies linked to carbon emissions? What planning techniques are being used to limit carbon production? It is rational to consider the answers to both of these questions concurrently, because strategic planning plays an important role in each case. To illustrate the importance and effectiveness of strategic planning for climate change, in concert with the principles of sustainable development and decision-making processes, specific examples are offered from both the public and private sectors.

The book *Creating a Climate for Change* discusses Santa Monica's comprehensive plan for sustainability.[2] This replicable model—developed in the 1990s and carried forward to today—is the result of progressive tradition, visionaries, champions, and community-wide support for a shared common good. It is a blueprint that links climate change with other environmental issues, economic development, and social equity in the larger context of the community's quality of life. Surrounded on three sides by Los Angeles County in California, the municipality of Santa Monica—just 21.5 km² (8.3 square miles), but with global vision—has consistently been at the leading edge of a movement for sustainable cities.

In 1999, Santa Monica was the first U.S. city to switch to 100% renewable electricity for all municipal facilities, which were also up-graded for energy efficiency. Of the city's public fleet 78% currently use reduced-emissions fuel. Since 2000, the Santa Monica Urban Runoff Recycling Facility (SMURRF) has reclaimed 95% of the water (much of it polluted) that previously flushed directly into the Pacific Ocean at the city's western edge.[3] Over the last decade, Santa Monica has developed a set of rigorous green building standards that are redefining the notion of healthy, efficient construction, and one being used as a model for other communities. Through its efforts since 1994, Santa Monica has shrunk the size of its ecological footprint, a measure of its resource use and waste sinks, by 5.7%. In a business-as-usual projection, Santa Monica's sustainable city program estimated that its GHG emissions would rise 14% above 1990 levels by 2010. Yet between 1990 and 2000, the city reduced its GHG emissions by 6%, a remarkable accomplishment. In June 2005 at the United Nations' World Environment Day in San Francisco, Santa Monica was recognized as the fifth most sustainable city in the United States.[4] Santa Monica's pre-dominant planning strategy was based on the Local Agenda 21—a framework developed at the 1992 United Nations Conference on Environment and Development—that called for "local authorities as the sphere of governance closest to the people—to consult with their communities and develop and implement a local plan for sustainability."[5] Santa Monica, with help from its elected visionary leaders and technical assistance provided by the Cities for Climate Protection (CCP), implemented such a plan. The city eventually updated its original sustainability plan from the 1990s in 2003, with support and participation by citizens and community groups. The guiding principles it incorporated in its revised sustainability plan include: stewardship for its natural environment, public/private partnership, a sustainability concept, socially-responsible values, interconnectedness, global vision, resource conservation, infrastructure modernization, and civic participation. Santa Monica also identified 70 sustainable indicators that are being measured for targeted performance.[6]

A prime conceptual framework used by private sector industrial and business enterprises is corporate social responsibility (CSR). The concept of CSR is based on the triad of economy, environment, and community, or the "triple bottom line" philosophy. A prime catalyst for action is potential cost savings that arise from making a business or industry more efficient. This has been demonstrated by a number of businesses in the United States that have reduced annual GHG emissions, while simultaneously reducing costs through savings in energy and materials consumption. For example, despite an almost 30% increase in production, by using 9% less total energy in 2002 than it did in 1990, DuPont reduced cumulative energy costs by $2 billion between 1990 and 2002.[7] Utilizing energy efficiency and conservation measures, IBM reduced costs by $791 million between 1990 and 2002. In 2002, BP announced that it had achieved its 10% reduction target (from 1990 levels), which utilized an internal emissions trading system, thus creating an estimated $650 million of value for the company.[8] In 2002 alone, Baxter International, a medical products and services company, reduced its generation of non-hazardous waste and use of packaging, saving the company $2.9 million.[9]

The cost savings achieved by these businesses illustrate opportunities that exist throughout the economy. Furthermore, in "learning by doing" they develop knowledge regarding methods, costs, and benefits that can be shared. Toyota, for example, has licensed patented hybrid technology to Ford for the production of Ford's Hybrid Escape SUV. This model reflects Ford's strategy to reduce GHG emissions from its automobiles by 2030 (Hakim, 2004).

Yet barriers remain to the exploration of similar opportunities. Without increased awareness or incentives, companies may not scrutinize their operations to find and develop these opportunities—a leverage point where better communication could make a difference. Fortune 100 firms have been surprised by the cost-saving opportunities available to reduce GHG emissions. These examples underscore an important planning technique: that environmental and social concerns, not just corporate or financial concerns, can be a true motivation for progressive business practices.

CAP AND TRADE PROGRAMS

In Chapters 10 and 13 of this book, information was provided on cap and trade programs regarding the "Industrial and Manufacturing Carbon Reduction Technologies" and "Financial Approaches to Carbon Reduction." However, it is important to address this subject in terms of a conceptual framework for reducing GHG emissions—from local, to regional, to national, to international perspectives.

Corporations, local and regional authorities, and NGOs are adapting a variety of actions to reduce GHG emissions (IPCC 2007). Corporate actions range from voluntary initiatives to emissions targets, and in a few cases, internal trading systems. The reasons corporations undertake independent actions include the desire to influence or pre-empt government action, to create financial value, and to differentiate a company's products and services. Actions by regional, state, provincial, and local governments include renewable energy portfolio standards, energy efficiency programs, emissions registries, and sectorial cap-and-trade mechanisms. These actions may influence national policies, address stakeholder concerns, create incentives for new industries and offer environmental co-benefits. NGOs promote programs that reduce emissions through public advocacy, litigation, and stakeholder dialog. The U.S. Congress considered legislation in 2008, known as the McCain-Lieberman Bill.[10] It is likely that such a program will be reconsidered with the support of the Obama administration.

A variety of policies, measures, instruments, and approaches are available to national governments to limit the emissions of GHGs; they include regulations and standards, taxes and charges, tradable permits, voluntary agreements, informational instruments, subsidies and incentives, research and development, and trade and development assistance. Tradable permit instruments include marketable permits and cap-and-trade systems. These instruments establish a limit on aggregate emissions by specified sources, require each source to hold permits equal to its actual emissions, and allow permits to be traded among sources.

CARBON EMISSIONS TRADING
MECHANISMS VS. EMISSIONS TAXES

These two topics were discussed separately in Chapter 13 of this book, under "Financial Approaches to Carbon Reduction." However, the objective here is to compare these two different, yet complementary, approaches that can result in successful carbon reduction programs in different sectors of the economy and levels of government.

Emission trading has become an important implementation mechanism for addressing climate change in many countries (IPCC 2007). The overall value of the global carbon market was over $10 billion in 2005, and in the first quarter of 2006, the transaction level reached $7.5 billion.[11] The EU developed the most advanced emission trading scheme (ETS). While the system is an international one, it bears many of the characteristics of a national program, with oversight by the European Commission, and a centralized regulatory and review mechanism. A larger global system of international trading is slowly developing through emissions credits generated by project-based mechanisms. Theoretically, a fully global ETS would provide market players and policy-makers with information thus far absent from decision-making: the actual, unfettered, global cost of GHG mitigation in a range of economic activities. In an international context, such a regime would mirror the information provided by national trading programs on a global scale.

An "emission tax" on GHG emissions requires individual emitters to pay a fee, charge, or tax, for every ton of GHG released into the atmosphere. An emitter must pay this per-unit tax or fee, regardless of how much emission reduction is being undertaken. Each emitter weighs the cost of emission control against the cost of emitting and paying the tax; the result is that polluters undertake to implement those emission reductions that are cheaper than paying the tax, but they do not implement those that are more expensive. Since every emitter faces a uniform tax on emissions per ton of GHG (if energy, equipment, and product markets are perfectly competitive), emitters will undertake the least expensive reductions throughout the economy, thereby equalizing the marginal cost of abatement (a condition for cost-effectiveness). Taxes and charges are commonly leveled on commodities that are closely related to emissions, such as energy or road use.

An emissions tax provides some assurance in terms of the marginal cost of pollution control, but it does not ensure a particular level of emissions. Therefore, it may be necessary to adjust the tax level to meet an internationally-agreed emissions commitment (depending on the structure of the international agreement). Over time, an emissions tax needs to be adjusted for changes in external circumstances, such as inflation, technological progress, and new emissions sources.[12] Innovation and invention generally have the opposite effect by reducing the cost of emissions reduction and increasing the levels of reductions implemented. If the tax is intended to achieve a given overall emissions limit, the tax rate will need to be periodically adjusted to offset the impact of new energy sources and emission mitigation technologies.

For clarification, a carbon tax is a levy on the carbon content of fossil fuels, providing economic disincentives for using fuels such as coal and oil. Since virtually all of the carbon in fossil fuels is ultimately emitted as CO_2, a carbon tax is equivalent to an emission tax on each unit of CO_2-equivalent emissions. Carbon taxes have the capability of being broadly applied to the carbon-based fuels across a county's entire economy. Emissions trading schemes can be designed to apply to the intra-company, domestic, and international levels and can focus on the primary emitters of greenhouse gases, especially fuels that are not related to transportation. The IPCC's Second Assessment Report adopted the convention of using permits for domestic trading systems and quotas for international trading systems.[13] Emissions trading under the Kyoto Protocol uses a tradable quota system, based on the assigned amounts calculated from the emissions reduction and limitation commitments.

Carbon taxes are most effective when imposed by central governments. Both Sweden and Norway imposed carbon taxes in 1991 to reduce greenhouse emissions. Along with other policies, the Swedish Ministry of Environment has estimated that their carbon tax has reduced emissions by an additional 20%.[14] The motor fuel tax is estimated to have reduced emissions from vehicles by 1.5 million metric tons in 2005. Norway's carbon taxes range from $16 to $63 per metric ton and apply to 68% of the country's total carbon footprint.[15] The government estimates that the carbon tax alone has reduced emissions by 15-20% since 1990.[16] Finland and Denmark, countries that have made major investments in renewables, also have carbon taxes in place. However, in order to successfully reduce carbon emissions, the funds from the tax revenues need to be directed toward financing substitute fuels and improving energy efficiency.

TRANSPORTATION SYSTEMS AND ALTERNATIVES

Settlement areas in the U.S. are expanding as a result of increasing population and economic growth; consequently, the use of private vehicles and related carbon emissions has increased.[17] Comprehension of the U.S. carbon footprint's geography and its relation to various transportation systems would significantly improve mitigation policies.[18] The following facts are important to understanding transportation systems in the U.S. in relation to carbon emissions:

1) Between 1982 and 2006, vehicle miles traveled (VMT) in the United States have increased by 47% per person, from an average of 10,950 km (6,800 miles) per year per capita, to almost 16,100 km (10,000 miles) per year per capita;

2) During the same period (1982 to 2006), national consumption of oil for transportation rose from 3.4 to 5.1 billion barrels per year, corresponding to nearly $500 billion;

3) U.S. GHGs from transportation represent 33% of total GHG emissions—automobiles and light trucks are the largest sources of GHGs;

4) People living in households within one-quarter mile of rail and one-tenth of a mile from a bus stop, drive approximately 7,100 km (4,400 miles) less annually, as compared to persons in similar households with no access to public transit; and

5) Communities that choose to invest in public transportation reduce the nation's carbon emission by 37 million metric tons annually—equivalent to the electricity used by 4.9 million households at an average of 10,000 Kwh/year/household.[19]

It is clear from the above facts that by reducing the growth in VMTs, easing congestion, and supporting more efficient land-use patterns, public transportation can reduce CO_2 emissions. These savings represent the beginning of public transportation's potential contribution to the national efforts to reduce CO_2 and GHG emissions and promote energy conservation. Moreover, the cost of fuel makes driving private vehicles even more prohibitive. Public transportation improvements have the potential to save U.S. households an average of $6,251 per year—perhaps more if the price of fuel increases.[20]

Higher density development—including transit-oriented development (TOD), multi-use buildings, and compact apartments and office space—is more energy efficient and extends public transportation's contribution by integrating it with other sectors of the economy, both in the United States and other urban centers of the world. In the "Public Transportation's Contribution to U.S. Greenhouse Gas Reduction" report, answers to the following four questions are provided: [21]

1) *How much net CO_2 is public transportation saving in the United States from the current levels of services being offered?*
   In 2005, public transportation reduced CO_2 emissions by 6.9 MMTs. If current public transportation riders were to use personal vehicles instead of transit, they would generate 16.2 MMTs of CO_2. Actual operation of public transit vehicles, however, resulted in only 12.3 MMTs of these emissions. In addition, 340 million gallons of gasoline were saved through transit's contribution to decreased congestion, which reduced CO_2e by another 3.0 MMTs. An additional 400,000 metric tons of GHG were also avoided, including sulfur hexafluoride, hydro-fluorocarbons (HFC), per-fluorocarbons, and CFCs.[22]

2) *How much additional CO_2 savings are possible, if incremental public transportation passenger ridership is increased?*
   A solo commuter switching his or her commute to existing public transportation in a single day, can reduce CO_2 emission by 7.5 kg (20 lbs.). An average private vehicle emission rate is about .37 kg (1.0 lb.) of CO_2 per mile. Over the course of a year, an individual could potentially reduce his or her CO_2 emissions by more than 1,791 kg (4,800 lbs), assuming 240 days of transit travel per year. This represents slightly more than two metric tons of CO_2 or about 10% of a typical two-car family household's annual carbon footprint.[23]

3) *What is the significance of using more public transportation at a household level and what households do to save additional CO_2?*
   The annual use of an automobile, driving an average of 19,300 km (12,000 miles) per year, emits 4.6 tons of CO_2 per year (one metric ton is equivalent to 2, 205 lbs). The carbon footprint of a U.S. household is about 22 metric tons per year. Reducing the daily use of one low-occupancy vehicle and using public transit can reduce a household's carbon footprint between 25% to 30%.[24]

4) *Are there favorable land-use impacts that public transportation contributes to that result in positive environmental and social benefits?*

The results range from a reduction in VMTs of between 2.35 km (1.4 miles) and 14.5 km (9 miles) for every transit passenger-mile traveled. The outcome would be more efficient use of roadways, less traffic, reduced road maintenance, parking facility optimization, and shorter highway-commute times.[25]

A similar analysis is being conducted with regards to travel by airplane, associated fuel consumption, carbon footprint, and measures to significantly reduce CO_2 and GHG emissions individually or in groups through new technologies, behavior changes, and smart aircraft/aviation systems management at all levels of the government and industry. Initiatives are underway. By 2012, airlines operating in or out of EU countries will be required to limit emissions to 97% of 2005 levels.[26] According to EU ministers, "the main objective is to reduce the impact of aviation on climate change, given the rapid growth of this sector."[27] In order to meet the reduction goal, airlines must either reduce their emissions or purchase carbon dioxide credits from other industries with surpluses.[28] One can also apply the lessons learned from public transportation's relationship with climate change, as described in this section, to other public mobility arenas such as ocean travel and cargo shipping.

SUMMARY

This chapter illustrates a myriad of carbon reduction programs as well as the public and private conditions necessary for their successful implementation. Carbon reduction strategies are closely linked with the perceived importance and relevance of sustainable development principles at all levels of society. Indeed, Agenda 21 encourages "local authorities as the sphere of governance closest to the people—to consult with their communities and develop and implement a local plan for sustainability."[29] The political viability of carbon reduction programs, such as cap and trade initiatives and emission taxes, is critical to motivating citizens and organization towards reform.

Another important factor in the successful implementation of carbon reduction programs is the decision-making processes involved. This aspect is often associated with economic and social factors. Moreover, a prime catalyst for action is the potential cost savings that arise from energy efficiency or long-term benefits associated with carbon reduction programs. Though carbon reduction programs seem expensive and complex, economic incentives such as an emission tax, a carbon market, or rising fuel prices make such programs economically attractive and in parallel with the profit-making emphasis of private industry and individual agents.

Carbon reduction programs are also supported through social and moral awareness about the related environmental benefits and overall improvements in the standard of living. Many forward-looking, visionary enterprises pursue carbon reduction programs through the umbrella of CSR initiatives, which incorporate a consideration for the economy, environment and community in a corporation's strategic planning. Corporations undertake such progressive measures for several reasons: to influence or preempt government action, to create financial value, and to differentiate the company's products and services. CSR and the propagation of grass roots movements, and non-profits that focus on sustainable development, fuel the realization of carbon reduction programs. Without increased awareness or incentives, companies may not scrutinize certain aspects of their operations to find and develop these opportunities.

Altogether, in order for a carbon reduction program to be truly successful and sustainable it requires political, economic, and social reinforcement. Such programs must be cost-effective and appeal to a wide range of individuals. Transportation systems such as auto and air travel have an important role to play. Consumers must invest in energy-saving technologies or practices, CEOs of corporations must have a visionary perspective, and political actors must encourage such programs for the benefit of their countries and future generations. Carbon reduction programs are not necessary altruistic, but the majority have long-term benefits and require both vision and advanced planning.

Endnotes

1. Roland, E. and Molina, M. (2001). *The CFC ozone puzzle: Environmental science in the global arena*. Presented at the First National Conference on Science, Policy and the Environment. Washington D. C.
2. Dilling, L. and Moser, S. (2007). *Creating climate for change*. Cambridge University Press.
3. Santa Monica Urban Runoff Recycling Facility (2007). City of Santa Monica. http://www.smgov.net/epwm/smurrf/smurrf.html, accessed 29 May 2008.
4. City of Santa Monica. (2005, 2 June). *Santa Monica honored as sustainability leader in U.S.* http://www.smgov.net/news/releases/archive/2005/epwm20050602.htm, accessed 29 May 2008.
5. UN Conference on Environment and Development (1992, June). *The Earth Summit*. http://www.un.org/geninfo/bp/enviro.html, accessed 2 June 2008.

6. City of Santa Monica (2008). *Sustainable Santa Monica.* http://www.smgov.net/epd/scp/index.htm, accessed 29 May 2008.

7. Dilling, L. & Moser, S. (2007). *Creating climate for change.* Cambridge University Press, p. 323.

8. *Ibid.*

9. *Ibid.*

10. Pew Center on Global Climate Change (2003). *Summary of the Lieberman-McCain climate stewardship act.* http://www.pewclimate.org/policy_center/analyses/s_139_summary.cfm, accessed 29 May 2008.

11. Intergovernmental Panel on Climate Change (2007, November). Report on *Mitigation of climate change.* International Emission Trading, p. 778. http://www.ipcc.ch/ipccreports/ar4-wg3.htm.

12. Tietenberg, T. (2000). *Environmental and natural resource economics.* New York: Harper-Collins Publishers.

13. Intergovernmental Panel on Climate Change (1995). *Second assessment report.*

14. David Suzuki Foundation (2008). *Carbon tax backgrounder—B.C. budget 2008.* http://www.davidsuzuki.org/files/climate/Briefing_Note_-_BC_Budget_2008.pdf, accessed 2 January 2009.

15. *Ibid.*

16. *Ibid.*

17. Brown, M. (2008, May). *Shrinking the carbon footprint of metropolitan America.* Brookings Institution Metropolitan Policy Program.

18. *Ibid.*

19. American Public Transportation Association (2008, March). *Fact sheet/transit news.* http://www.apta.com/, accessed 29 May 2008.

20. Bailey, L. (2007, January). *Public transportation and petroleum savings in the U.S.: reducing dependence on oil.* ICF International. http://www.apta.com/research/info/online/documents/apta_public_transportation_fuel_savings_final_010807.pdf, accessed 29 May 2008.

21. Energy Solutions Operation (2007, September). *Public transportation's contribution to U.S. greenhouse gas reduction.* Science Applications International Corporation.

22. *Ibid.*

23. *Ibid.*

24. *Ibid.*

25. *Ibid.*

26. Airlines to limit emissions in Europe (2008, November 3). *The Courier-Journal.* p. E3.

27. *Ibid.*

28. *Ibid.*

29. UN Conference on Environment and Development (1992, June). *The Earth Summit.* http://www.un.org/geninfo/bp/enviro.html, accessed 2 June 2008.

Chapter 15

The Future of
Carbon Reduction Programs

"The 800-pound environmental gorilla that is carbon... It's an issue that we've been analyzing as a company for several months. The first point I want to make is that I firmly believe that when we talk about carbon and its impact on global warming, we are not talking about just another business issue. We are talking about the type of business issue that comes along, perhaps only once in a century.

Make no bones that we at NRG Energy want to have a seat at the table in the carbon policy and I will be equally frank in telling you that right now despite our best efforts as a company I am pessimistic and my pessimism is rooted in my extreme dismay with the carbon position of the power industry as a whole. While some forward thinking power industry executives have taken a stand on carbon, the broader utility industry view as represented by the official position of the utility trade association can best be described as—see how this works: "See no carbon, hear no carbon, and speak no carbon." More specifically for those of you who do not know the official utility industry view is that carbon restraints should be entirely voluntary. A carbon position based on voluntary restraints to me is unwise and ultimately self defeating because it's increasingly out of touch with the rapidly hardening position of main stream America on the issues of carbon emissions and global warming and what will be even more infuriating to the public is the blazing cynicism of the power industry's position—advocating reliance solely on voluntary restraints at a time when the American power industry itself is proposing to construct nearly 100 gigawatts of new coal-fired generation. Note that if all the traditional coal plants under development across the United States are built, we will be adding an incremental 700 million tons of carbon, of CO_2 emissions every year, an incremental amount of carbon equal to the total carbon emissions of Spain and France combined."

Remarks by David Crane, CEO of NRG Energy (an
independent power producer) to the Merrill Lynch Power &
Gas Leaders Conference on Sept. 26, 2006[1]

REDUCING CARBON EMISSIONS

U.S. anthropogenic emissions of greenhouse gases have increased 16.9% from 6,113 million metric tons in 1990 to 7,147 million metric tons in 2005.[2] From 2005 to 2007 GHG emissions jumped another 1.9 percent, rising to 7,282 million metric tons.[3] Forecasts of carbon emissions are equally alarming. By 2007, U.S. energy-related CO_2 emissions had increased to 5,984 million metric tons. Recent changes in GHG and carbon emissions correlate with growth in electricity demand because much of U.S. electrical production relies on carbon-based fuels. Not surprisingly, U.S. electricity demand is growing by 1.5% annually. Though this growth is likely to be interrupted by the recession of 2008-2009, by 2030, electrical demand is projected to increase 50%.[4] At a time when multinational oil companies are having difficulty gaining access to new oil resources, the international Energy Agency (IEA) projects that the world's demand for oil will increase from 85 million barrels daily to 106 million barrels by 2030.[5] Unless a greater portion of this growth is accommodated by alternative fuel sources, carbon emissions are likely to increase. Carbon emissions will also be impacted by further economic development and population growth in developing countries. Some estimates indicate that annual carbon emissions may double to 80 billion metric tons.[6]

What can be done to thwart environmental problems associated with CO_2 gas emissions? The "low-hanging fruit" of carbon offsets has been addressed in many countries by the Kyoto Protocol.[7] This includes the potent greenhouse gas HFC23 which is lucrative in carbon markets since each ton is equivalent to 11,700 tons of carbon dioxide.[8] The other potent greenhouse gases are nitrous oxide and methane. Reducing the emissions of these gases is critical to the success of our carbon reduction efforts. Implementing economically-feasible technologies is the next step. Low-carbon energy sources need to expand from 19% of the world's energy mix in 2006 to 26% by 2030.[9] This will require an estimated investment of $4.1 trillion in infrastructure and equipment including more energy-efficient automobiles, appliances and buildings.[10] These improvements will yield fuel cost savings which have been estimated to be greater than $7 trillion.[11]

The rare potential for breakthrough technologies can be both inspiring and burdensome, especially when many commercially available technologies remain to be implemented. For example, some argue that the last quantum leap in energy technology is the development of nuclear energy. This development occurred over 60 years ago. Still, the use of nuclear energy to generate electricity has unresolved waste-disposal issues. It also has limited political feasibility. The development of nuclear power is expensive, and is often linked to concerns about the proliferation of nuclear weapons.

Other promising technological advancements have also occurred in recent years. Carbon capture and sequestering provide examples of technologies that will need extensive research before becoming commercially scalable. Though costly, the use of liquid sodium hydroxide (lye) holds promise as it has the ability to absorb and capture CO_2 from airstreams. It is most feasible for locations where emissions are concentrated and are being exhausted, such as the stacks of coal-fired power plants. CO_2 released into the atmosphere can be captured and stored in oceans, caves, abandoned oil and natural gas fields, and elsewhere. Not all of these sequestration techniques are environment-friendly. When stored in water, carbonic acid may form and disrupt marine life. In contrast, planting trees and other forms of vegetation is considered one of the most cost-effective sequestration techniques. This solution improves the natural system's capacity not only to store carbon, but also to generate oxygen. Finally, industrial uses for CO_2 will reduce the need for absorption and sequestration. CO_2 can be combined with catalysts to create substitutes for other products. One use may be to create new forms of biodegradable plastics to make containers for soft drinks.

Renewable energy will be an important part of our portfolio of sustainable energy solutions. Renewable technologies such as wind power, solar and waste-to-energy plants, are available today to reduce energy consumption and cut dependency on fossil fuels. During 2008 in the U.S., wind energy added a record 8,300 megawatts of capacity while only five new coal plants generating 1,400 megawatts came on line.[12] These developments offer economic benefits through job creation and cost reductions. Waste-to-energy plants as one example, already convert about one-eighth of U.S. landfill wastes to energy. Yet sustainable energy development has its own "upside-down" economics. In most cases, economic trade-offs are necessary. While solar and wind power often have higher initial costs, they also have negligible carbon emissions but these are related to primarily to equipment manufacture. Conventional energy solutions have lower initial costs, burdensome carbon emissions, and rely on costly fuels. Furthermore, as the right to emit atmospheric carbon becomes more expensive, alternative energy solutions will become more viable.

INFORMATION TOOLS FOR SUSTAINABLE SOLUTIONS

Information tools or "information instruments" refer to various labeling programs, as well as rating and certification systems. Article 6 of the United Nations Framework Convention on Climate Change (UNFCCC) on Education, Training, and Public Awareness calls on governments to promote educational and public awareness programs, public access to information, and public participation.[13] Information tools enable such public disclosures, typically by an industry to its customers, which relate a particular service or product to its environmental impact. Public disclosures may positively affect the environment by allowing consumers to make better-informed decisions.

Information tools also serve to steer and support various government policies. The literature on climate change addresses a wide variety of national policies and measures, intended to limit or reduce GHG and CO_2 emissions. According to the IEA, all the governments of the world should establish and maintain a comprehensive framework to monitor energy consumption trends at the end-use level, and support work that addresses gaps in statistical databases. This includes regulations and standards, taxes and charges, tradable permits, voluntary agreements, subsidies, financial incentives, research and development programs, and information instruments. Moreover, policies addressing trade, foreign direct investment and social development goals, also affect CO_2 and GHG emissions. Climate change policies, when integrated with a government's agenda, are components of sustainable development policies. Sustainable development is commonly defined as "development that meets the requirements of present generations, without compromising the ability of future generations to meet their own needs."[14] A three-pronged approach to sustainability involves environmental stewardship, economic prosperity, and social justice.

Significant emission reductions have occurred as a result of government action to address energy security and other national needs. Brazil's development of a transport fleet driven by bio-fuels is an example of a sustainable development policy that provides both economic benefits and reductions in CO_2 emissions. Moreover, these policies can be tailored to fit each nation's natural resources, population needs, technological capacity, and infrastructure.[15] The advantages and disadvantages of any given policy will vary with each country's circumstances. The IPCC has identified four main criteria that are widely used by policy-makers to select and evaluate policies:

- Environmental Effectiveness
- Cost Effectiveness
- Distributional Effects
- Institutional Feasibility[16]

These criteria indicate that in order for climate change mitigation policies to be successful, the actions taken by national governments must also be supported by the private sector and civil organizations (e.g., non-governmental organizations or NGOs). The coordination of broad-based carbon reduction policies involves many aspects of society and requires the appropriate information tools to gather and sustain support. The public and private sectors are interlinked in the struggle towards a sustainable society.

REDUCING CARBON EMISSIONS—FUTURE STRATEGIES

Concerted global action is needed to address our daunting energy challenges. Without urgent action, energy and economic security will deteriorate and climate change will become a devastating reality. In 2005 the IEA was asked to advise the G8 nations (France, Germany, Italy, Japan, the United Kingdom, Russia, Canada and the United States) on alternative scenarios and strategies for a "clean, clever, competitive energy future."[17] From 2005 to 2008, the IEA undertook an expansive assessment of available options, identifying the steps needed to implement such strategies. According to the 2008 IEA report to the G8 Summit, the world's energy challenges frequently correlate with economic growth.

Hence, using the development of industrialized countries as a paradigm for developing nations to follow is foolish. The IEA sought an alternative approach to economic growth and development. It's most significant recommendation was *Energy Revolution Now*, which details a strategy emphasizing the need for energy supplies that are clean, secure, and contribute to economic growth. Energy policies that do not balance these objectives will likely prove unsustainable.

The path to sustainability begins with improving energy efficiency throughout the global economy. It has been said that "energy efficiency is the cheapest form of power, and with aggressive efforts we can achieve dramatic gains."[18] Energy efficiency measures have the greatest potential to procure immediate CO_2 reductions at low cost. The IEA proposed a set of policy recommendations for promoting energy efficiency. This proposal could reduce global CO_2 emissions by 8.2 gigatons per year by 2030. What's more, the industrial sector's application of proven technologies and best practices on a global scale could save 18 to 26% of its energy use. The IEA-recommended "buildings-related measures" could deliver global savings of 1.4 gigatons of CO_2 annually by 2030. This figure is larger than Japan's annual CO_2 emissions in 2005. IEA recommendations in "equipment and lighting" would save another 2.2 gigatons per year. Further potential for emission reductions are found in the transportation sector. The 2005 level of per-kilometer energy use for light-duty vehicles could be cut in half. This step would utilize every available fuel-efficient technology, including engine/drive-train improvements, aerodynamics, weight reduction, low rolling-resistance tires, and other energy-efficient components.

To promote energy efficiency policies, the IEA has developed energy indicators to analyze the interaction among economic and human activity, energy use, and CO_2 emissions. These energy indicators are essential tools for measuring national and international performance and for analyzing trends.

The following is a summary of energy efficiency recommendations prepared by the IEA for the G8, under the Gleneagles Plan of Action. It consists of 25 separate qualitative and quantitative steps, as shown in Table 15-1.

Table 15-1. International Energy Agency Consolidated Energy Efficiency Recommendations

- Measures for increasing investment in energy efficiency
- National energy efficiency strategies and goals
- Compliance, monitoring, enforcement, and evaluation of energy efficiency measures
- Energy efficiency indicators
- Monitoring and reporting progress with the IEA energy efficiency recommendations themselves
- Building codes for new buildings
- Passive energy houses and zero energy buildings
- Policy packages to promote energy efficiency in existing buildings
- Building certification schemes
- Energy efficiency improvements in windows
- Mandatory energy performance requirements or labels

- Low-power modes, including standby power, for electronic and networked equipment
- Televisions and 'set-top' boxes
- Energy performance test standards and measurement protocols
- Best practice lighting and the phase-out of incandescent bulbs
- Ensuring least-cost lighting in non-residential buildings and the phase-out of inefficient fuel-based lighting
- Fuel-efficient tires
- Mandatory fuel efficiency standards for light-duty vehicles
- Fuel economy of heavy-duty vehicles
- Eco-driving
- Collection of high-quality energy efficiency data for industry
- Energy performance of electric motors
- Assistance in developing energy management capability
- Policy packages to promote energy efficiency in small- and medium-sized enterprises, businesses, and organizations
- Utility end-use energy efficiency schemes, programs, and initiatives

The U.S. federal government and the state of California, along with many large and small cities within the U.S., are adopting many of the IEA recommendations. The IEA's influence is evident in the various regulations, implementation guidelines, programs, projects, and innovative climate change delivery systems being adopted by cities.

MERGING LOCAL, STATE, AND FEDERAL GOVERNMENT POLICIES

Since the Industrial Revolution in the mid-1800s, energy usage has dramatically increased. Furthermore, a 2005 report by the Energy Information Administration indicated that primary energy consumption in the U.S. has increased from 75.7 to 100.3 quadrillion Btus over the past 30 years.[19] Our use of energy has created higher standards in production and delivery, improvements in health care, broader accessibility to transportation and communication. Regrettably, energy use has also led to unanticipated negative consequences. Some examples are the development of critical energy dependencies, geopolitical struggles over resources, and excessive air pollution from CO_2 and GHG emissions.[20]

Many major U.S. policies that address energy efficiency were enacted after the oil supply disruptions of the 1970s. Throughout the oil embargo, citizens in the U.S. experienced price shocks and energy shortages. Interest in energy efficiency, as a way to reduce dependency on foreign oil, initiated at the state level. Since then, state efforts to mitigate climate change have triggered recognition from federal initiatives as well as business communities. Multi-state alliances and partnerships with NGOs have created substantive political momentum. State influence stems, in part, from a "patchwork" quilt of state and local policies. Variance in state legislation that deals with energy efficiency can cause difficulties in interstate commerce.

It is important that initiatives and policies created at the state level result in actual CO_2 and GHG emissions reductions. The greatest reduction results are typically achieved by states adjusting their own operations. State action may also impact the actions of cities and towns (e.g., local governments). For example, approximately 60 communities in the northeastern U.S. have joined the Cities for Climate Protection Campaign. The ICLEI (International Council for Local Environmental Initiatives, now known as Local Governments for Sustainability) opened an office in the region to coordinate municipal climate action. Hundreds of major U.S. cities have also joined initiatives under the leadership of the U.S. Conference of Mayors. This partnership offers local assistance, policy guidance, and technical support to municipalities dedicated to reducing their CO_2 and GHG emissions.

Altogether, a wide variety of policy instruments have developed at the local, state and national level, making a comprehensive energy policy difficult to achieve. Still, "leadership by example" is the explicit goal of all levels of the government—local, state, national—and each continues to make progress in energy and environment issues. Land-use regulation, local growth strategies, and public transportation are the responsibilities of state and local governments. These strategies benefit from carefully-planned support from the national government. Integrated policy approaches that bridge the inconsistencies between local, state, and national governmental entities will reduce CO_2 and GHG emissions.

CARBON MARKETS

The quest to reduce carbon emissions is creating both challenges and opportunities. Renewable energy will play a key role, as will efficiency improvements in energy-consuming systems. Thus, the global market for renewable energy and related technologies is expected to grow to $167 billion by 2015.[21] Investments in alternative energy and energy efficiency will increase as carbon trading becomes firmly established. Carbon trading introduces market reforms that will hopefully reduce carbon emissions. Market theorists believe that such reforms depend on the price of carbon emissions. Moreover, as GHG emissions become more expensive, industry decision-makers will opt to reduce their industrial emissions rather than continuing to purchase emission credits.[22]

The European Climate Exchange (ECX) began in 2005 and is the primary exchange operating in the European Union Emissions Trading Scheme. It includes over 65 global companies and many other corporations and institutions. The International Transaction Log interlinks United Nations initiatives with those of Japan, New Zealand and 12 European countries.[23] This enables EU factories and power stations to use carbon credits managed by the UN under the Kyoto Protocol, in addition to those managed by their national governments. Since 2005, 12,000 industrial plants in the EU have been able to buy and sell rights to release carbon into the atmosphere.[24]

The EU Emissions Trading Scheme is a way that companies exceeding individual CO_2 emissions targets can purchase allowances from companies that have exceeded their targets.[25] This program is an attempt to cap carbon emissions on an economy-wide basis. It includes industries such as power generation, iron and steel production, glass, cement, pottery and masonry product manufacturing.[26] Trading has been surprisingly active. Over 780 million European allowances have been traded.[27] Not surprisingly, emissions are now consistently factored into the operations of installations throughout the EU.

The CCX is a voluntary carbon-trading market and the first legally binding GHG emissions allowance trading system in the U.S. Participants in the trading system include members of industry (DuPont, United Technologies, Ford), local governments (Miami-Dade County, City of Chicago), corporations (Knoll, Steelcase, Safeway), institutions (Michigan State University, University of Oklahoma, Hadlow College), and other entities. The carbon exchange market encompasses two phases. Phase I (2003-2006) was a period in which members committed to reduce emissions 4% below their baseline. In Phase II (2007-2010) all CCX members committed to reduction schedules that would reduce emissions by 6%.

The commodity traded is a carbon financial instrument (CFI), a legally binding contract to meet carbon reduction targets. Each CFI contract represents the equivalent of 100 metric tons of CO_2. Table 15-2 illustrates how U.S. markets are pricing carbon:

According to investor Sunil Paul, "there are huge forces at work right now. With the subsidies that are already in place in California, and markets for carbon credits emerging, you have the perfect conditions for innovative companies to capture a piece of the $1 trillion U.S. electricity market."[28]

In the first quarter of 2008, the New York Mercantile Exchange provided a venue for the trading credits that offset GHG emissions. The partners in the exchange are NYMEX, Evolution Markets (a broker of environmental credits), and investment banks.[29] The exchange hopes to resolve concerns over the credit-worthiness of buyers and sellers and provide third party guarantees of the quality of the carbon credits. The exchange will also handle products such as carbon reduction allowances, voluntary U.S. pollution offsets and renewable energy credits that will be used to subsidize alternative energy projects. The U.S. market for voluntary carbon credits is roughly $100 million per year. It is dwarfed

Table 15-2. Pricing History of CCX Carbon Financial Instruments

Source: Chicago Climate Exchange. http://www.chicagoclimatex.com/content.jsf?id=821, accessed 1 October 2007.

by the $60 billion worldwide market for carbon credits.[30] This is due to the fact that most world markets are regulated.

THE FUTURE OF ENERGY SUBSTITUTES

There are alternative ways to produce energy that address both fuel production and carbon capture. Some fuel substitutes are only 50% oil; these can be processed into biodiesel fuels and jet fuels. However, such processes can cost over $5 per liter ($20 per gallon) and get very expensive.[31] One such example is found in the conversion of algae oil into biodiesel. In 2006, a company in New Zealand demonstrated a Range Rover powered by an algae biodiesel blend. Still, it may be years before the fuel is commercially available.[32] Despite cost, algae technologies are a potentially scalable and marketable alternative to fossil fuels. While a hectare of corn yields only 120 liters (12.8 gallons per acre) of oil annually, a hectare-sized pond of algae can produce approximately 40,200 liters (4,000 gallons per acre) of oil.

Another foreseeable benefit in algae technology is evidenced in the organism's resilience and adaptability. Algae can be grown almost anywhere, even in seawater. It can be used to process sewage and pollutants from stationary sources such as power plants. A company in Boulder, Colorado, A2BE Carbon Capture, is developing a unique photo-bioreactor machine for growing and harvesting algae, which consumes CO_2 and NO_x from flue gas emissions. This technology would allow for on-site extraction and continuous production of bio-fuel compounds, protein, fertilizer and methane gas. James Sears, the company's president, says, "It's complex, it's difficult and it's going to take a lot of players" to commercialize the technology.[33] Yet he remains optimistic about the use of algae.

The transportation sector is also being targeted by mitigation technologies and strategies. Transportation systems are significant contributors to CO_2 and GHG emissions. The most promising short-term strategy for the transportation sector is continued incremental improvements in vehicle design. Great opportunities for CO_2 and GHG emissions reduction are found in electric-drive technologies, including hybrid-electric power trains, fuel cells, and battery electric vehicles. The use of alternative fuels, in combination with improved conventional and advanced technologies, provides potential for even larger reductions.

The transportation sector plays a crucial and growing role in world energy use and the consequential emissions of CO_2 and GHG. In 2004, transportation accounted for 26% of the world's total energy use, generating about 23% of energy-related CO_2 and GHG emissions. The growth rate of energy consumption (from 1990-2002) was highest in the transportation sector. Virtually all transportation energy comes from oil-based fuels—largely diesel (about 31%) and gasoline (about 47%). One consequence of this dependence is that the CO_2 emissions from the different transportation sub-sectors are roughly proportional to their energy use.[34] There is little doubt that transportation activity will continue to grow at a rapid pace. However, the shape of the demand and the means by which it is satisfied depends on several factors:

1) It is not clear whether oil can continue to be the dominant energy source for transportation. Alternative fuels include heavy oil, oil sands, oil shale, natural gas, coal, and biomass. Other options are gaseous fuels like natural gas or hydrogen, and electricity, with both hydrogen and electricity capable of being produced from a variety of feedstocks. However, all of these alternatives are costly, and several—especially liquids from fossil resources—increase CO_2 emissions significantly unless additional carbon sequestration techniques are in place.

2) The growth rate and shape of economic development—the primary driver of transportation demand—is uncertain. If China and India, as well as other Asian countries, continue to rapidly industrialize and Latin America and Africa fulfill much of their economic potential, transportation demand will grow rapidly. Even in the most conservative economic scenarios, considerable increases in transportation are likely.

3) Transportation technology has been evolving. For example, although hybrid electric-drive trains have made a strong early showing in the Japanese and U.S. markets, the ultimate market penetration will depend on future cost reductions. Other near-term options include using light-duty diesel from Europe. Long-term opportunities will require more advanced technology. Examples include the production of new biomass fuels (beyond those made from sugar cane in Brazil and corn in the U.S.), fuel cells that run on hydrogen, and battery-powered electric vehicles.

4) Current trends suggest a growing dependence on private cars; however, other alternatives exist and can be incorporated into urban planning processes. Some examples are cities like Curitiba and Bogota in Colombia, with their rapid bus transit systems. Also, non-motorized transportation (NMT), such as walking and cycling, can also help reduce CO_2 and GHG emissions. These methods of transportation are most effective when incorporated into a comprehensive urban transportation system network.

REDESIGNING BUILDINGS WITH CARBON REDUCTION PRACTICES

A range of opportunities exists for reducing energy use and related CO_2 and GHG emissions in buildings. These may be tackled incrementally or comprehensively, as budgets and time allow. In new construction projects, a common goal is to improve energy efficiency by 30%, 50% or 70% more than local building codes. Such targets are viewed as steps for maximizing design creativity. Some of the successful energy efficiency elements, components, and systems include:

- Building envelope—windows, doors, walls, roof, etc.
- Lighting—natural daylighting and connected loads
- Mechanical and HVAC systems
- Energy management and control system (EMCS)
- Electrical equipment and systems
- Hybrid systems like combined heat and power (CHP), distributed generation
- Alternative energy opportunities

In addition, there are numerous energy solutions that are outside of the energy professional's traditional realm.[35] Green building designers can be instrumental in bringing these solutions to market and finding their best applications. Examples include landscape techniques that provide shade and reduce cooling loads, new building orientation that maximizes or minimizes solar exposure or natural daylight, water conservation that decreases use and heating demands, the use of alternative fuels, leak detection and reparation (especially of steam lines), and the development of innovative management approaches and technologies for comfort and productivity of building occupants.

An important tool for developing green buildings and procuring climate change action, according to Rappaport and Creighton, is the professional community.[36] Advocates working in these areas recognize the importance of expertise and the use of comprehensive literature on energy efficiency and high-performance buildings. Conferences, training sessions, and trade magazines are important sources of current information. These are developed by organizations such as:

- American Society of Heating, Refrigerating, and Air Conditioning Engineers (ASHRAE) develops many industry standards and best practices;
- U.S. Green Building Council's Leadership in Energy and Environmental Design (LEED) program develops certification programs for green buildings and facilities;
- U.S. Environmental Protection Agency's Energy Star program which provides education and analysis tools; and
- The Association of Energy Engineers which sponsor education programs, training seminars and professional publications.

Efforts to improve facilities on educational campuses, such as projects that reduce the use of carbon-based energy, have demonstrated how comprehensive efforts can significantly reduce GHG emissions. The potential to save money is equally impressive. Actions to improve energy efficiency and building operations are underway at many institutions. Opportunities for improving efficiency, fuel switching, and alternative energy in new and existing buildings are abundant. In many design and renovation projects, the largest savings are realized by finding opportunities for optimizing building systems including lighting, HVAC, controls, and thermal envelope. To consider buildings as integrated systems rather than a collection of unrelated subsystems represents a new paradigm in approaches to design solutions. The emergence of energy efficient, high performance, "green" buildings will continue to require a systems approach to the design and use of innovative technologies.

ADVANCES IN ELECTRICAL UTILITY SYSTEMS

Utilities will play a key role in reducing carbon emissions by shifting priorities and reconsidering their electrical generation projects. Green power production is becoming their preferred approach. Nevada's LS Power has postponed construction of a 1,600 megawatt coal-fired power plant and will instead "focus on tapping the state's geothermal, wind and solar potential."[37] Interstate Power in Iowa is pursuing construction of a 200 megawatt wind farm. Electric Co-ops abandoned the idea of constructing a coal-fired plant in Montana and are instead developing a plant fueled by natural gas.[38]

Utility control technologies have the potential to reduce electricity demand and thus reduce the need for more base-load power plants. Smart electric metering technologies can provide sophisticated demand response capabilities, using highly-automated digital pathways to communicate data. They offer high-speed data transmission from control centers to individual electric meters—even those for residences.

Using transmission systems with wireless communications and utility interconnections, "smart grid" controls can off-load energy-consuming equipment such as washers, dryers and water-heaters. This technique lowers the electrical demand in real time as needed. The appliances have onboard digital decision-making capability and communicate with each building's smart meters, making decisions as to when the appliances should operate. This technology is especially useful during peak-load conditions. Refrigerators will choose the best time to run their compressors to make ice. Dishwashers and clothes dryers will choose the best periods to operate their cycles. A home's water heater and thermostat (in communication with occupancy sensors) will make decisions as to when to change temperature set points, optimizing occupancy conditions and minimizing energy use. Any electrical generation equipment installed on-site, such as solar photovoltaic or wind generation equipment, can be metered and priced in real-time as needed by system demand.[39] Such technologies can provide demand response, real-time pricing, remote control and can use any battery storage that may be available, such as the batteries in the electric vehicle parked in the garage.

The world's first smart grid, which incorporates many of these technologies, is being constructed in Boulder, Colorado. The $100 million pilot project will test the capabilities of smart grid applications—50,000 customers will be served with automated net-metering control systems.[40] This system will be able to "predict and prevent many outages, communicate with smart appliances and thermostats to avoid overloads and peak costs, give customers real-time pricing... and integrate plug-in electric hybrids as potential backup generators or battery storage available to the grid."[41] This will result in a reduced carbon footprint, greater consumer control and lower costs for the local utility.

SATISFYING A CONFLUENCE OF IMPERATIVES

We face a confluence of global imperatives regarding climate change. We are experiencing an unsustainable trajectory of atmospheric CO_2 concentrations. While we must grow our economies, we must also protect them by reducing atmospheric carbon emissions and also reducing our dependence on carbon-based fuels. The greatest research and development challenge for the near term does not involve securing domestic energy supplies of fossil fuels—there is a broad resource base of fossil fuels to meet most current supply requirements.[42] The availability of many of these resources will diminish over time and substitutes will be necessary. The problem for the 21st century is how we can "provide clean, reliable, affordable energy that reduces CO_2 emissions."[43] Our inability to precisely predict the impacts of climate change further complicates our situation. If natural systems become incapable of storing CO_2 in the quantities necessary to prevent additional changes to our ecosystems, the only alternatives available to us are to advance and implement technologies to alleviate GHG emissions. There are tradeoffs to consider in our efforts to mitigate climate change and reduce carbon emissions. Since it is invariably less expensive to reduce energy use than to create new energy, rational approaches begin with implementing energy saving and efficiency improvements.[44] This will continue to be a fundamental cornerstone of energy policies. There is evidence that this is being widely recognized by the U.S. government. On February 17th, 2009, the economic stimulus package signed by President Barack Obama included $16.8 billion to expand renewable energy and efficiency programs.[45]

U.S. greenhouse gas regulations and policies are in flux. In April 2009, the U.S. EPA released a proposed finding on CO_2 and five additional greenhouse gases (methane, nitrous oxide, hydrofluorocarbons, perfluorocarbons, and sulfur hexafluoride). The agency stated that "in both magnitude and probability, climate change is an enormous problem" and that greenhouse gases "endanger public health and welfare within the meaning of the Clean Air Act."[46] It cited man-made pollution as a "compelling and overwhelming" cause for global warming. According to EPA Administrator Lisa Jackson, the finding "confirms that greenhouse gas pollution is a serious problem now and for future generations."[47] This finding was prompted by a Supreme Court decision in April 2007 ruling that greenhouse gases are pollutants as classified by the Clean Air Act and regulation is required if human health is threatened.[48] A final finding is forthcoming as is congressional action. House Speaker Nancy Pelosi has stated that, "the Congress is working on a comprehensive solution to global warming, and I am committed to moving clean energy legislation this year that will include perspectives from across our nation to create jobs, improve our national security, and reduce global warming."[49]

This redirection of U.S. policy is already impacting the utility industry's plans to increase coal-fired electrical power generation. AES Corporation recently withdrew an application to construct a 600 MW coal-fired power plant in Oklahoma as part of a "broader strategy to reevaluate" their growth plans.[50] In Louisiana, three coal-fired power plants or expansions are undergoing state review. Louisiana Generating was recently sued by the EPA for failing to install modern pollution control equipment when its Big Cajun 2 Plant underwent major modifications.[51] Michigan has placed seven power plants on hold while Governor Jennifer Granholm attempts to shift her state to power from cleaner, more sustainable energy sources.[52] In Kansas, a state with a successful wind power development program, Governor Kathleen Sebelius has vetoed three legislative attempts to approve two large coal plants proposed by Sunflower Electric.[53]

The success of carbon reduction policies and programs will be tested by their ability to actually cause reductions in the levels of carbon and greenhouse gases in the earth's ecosystems. While regional carbon markets and local incentives offer encouraging models, in order to be successful, a concerted world-wide effort will be necessary. A common mission must be adopted that includes policy integration, program development and the implementation of appropriate technologies. This creates the opportunity for the creation of a new world energy order which will involve research, decentralized energy solutions, energy efficiency measures, and greater use of alternative energy and carbon reduction technologies. This energy transformation will require a stronger sense of participation and greater investments by our international, regional and local governments. The policies and programs that these governments develop and implement will be important. This transformation will also require us to reevaluate and change our traditional behaviors and reconsider our individual decisions.

Endnotes

1. Clean Energy Partnership. For full text see http://www.cleanenergypartnership. org/news/article_detail.cfm?id=231, accessed 21 September 2007.
2. Energy Information Administration (2005). *Emissions of greenhouse gases in the United States, Executive summary.* www.eia.doe.gov/oiaf/1605/ggrpt/pdf/executive_summary.pdf, accessed 30 July 2006.
3. Energy Information Administration (2008, December 3). *Emissions of greenhouse gases report.* http://www.eia.doe.gov/oiaf/1605/ggrpt, accessed 24 December 2008.
4. *Power execs: cap carbon emissions* (2006, October 22). http://www.wired.com/techbiz/media/news/2006/10/71990, accessed 22 September 2007.
5. Porretto, J. and Wardell, J. (2008, November 13). Agency warns of energy crisis. *The Courier-Journal.* p. A1-A2.
6. *Power execs: cap carbon emissions* (2006, October 22). http://www.wired.com/techbiz/media/news/2006/10/71990, accessed 22 September 2007.
7. Carbon Positive (2006, September 13). *Kyoto's CDM approaches the crossroads.* www.carbonpositive.net/viewarticle?articleID=432, accessed 8 January 2009.
8. *Ibid.*
9. Porretto, J. and Wardell, J. (2008, November 13). Agency warns of energy crisis. *The Courier-Journal.* p. A2.
10. *Ibid.*
11. *Ibid.*
12. Brown, M. (2009, March 8). Foes and costs halt coal plant expansions. *The Courier-Journal.* p. A4.
13. United Nations Framework Convention on Climate Change (2008, January 24). *Education and outreach.* http://unfccc.int/cooperation_and_support/education_and_outreach/items/2529.php, accessed 17 November 2008.
14. GreenUP. http://www.greenupdays.org, accessed 2 January 2009.
15. Intergovernmental Panel on Climate Change (2007). Report on *Mitigation of climate change.* p. 341-342.
16. *Ibid.*
17. IPCC (2007). Report on *Mitigation of climate change.* p. 257.
18. Bailey, J. (2009, January 5). Get the energy future right. *The Courier-Journal.* p. D1.
19. See Energy Information Administration website for historical energy information in the U.S. http://www.eia.doe.gov/oiaf/kyoto/pdf/fd09abv.pdf, accessed 10 January 2009.
20. Moser, L. and Dilling, S. (2007). *Creating a climate for change.* Cambridge University Press.
21. Copeland, M. and McNichol, T. (2006, November). Here comes the sun. *Business 2.0.* p. 100.
22. Hotz, R. (2007, February 11). Wanted: carbon emissions. *The Courier-Journal.* p. A11.
23. Carr, M. (2007, August 30). EU, UN emissions systems to be linked by November 17. *Bloomberg.* http://www.bloomberg.com/apps/news?pid=newsarchive&sid=a GZeMMQ1Gok0, accessed 1 October 2007.
24. EurActiv (2007, June 29). *EU emissions trading scheme.* http://www.euractiv.com/en/sustainability/eu-emissions-trading-scheme/article-133629, accessed 10 July 2007.
25. *Ibid.*
26. *Ibid.*
27. International Emissions Trading Association (2006, October 13). *IETA position paper on EU ETS market functioning.* http://www.ieta.org/ieta/www/pages/getfile.

php?docID=1926, accessed 9 July 2006.
28. Copeland, M. and McNichol, T. (2006, November). Here comes the sun. *Business 2.0.* p. 100.
29. Davidson, P. (2007, December 13). Green Exchange takes root soon. *USA Today.* p. 3B.
30. *Ibid.*
31. Karanowski, S. (2007, December 9). Algae fueling new renewable energy research. *The Courier-Journal.* p. D1.
32. *Ibid.*
33. Karanowski, S. (2007, November 29). There's oil in that slime. *Associated Press.* http://ap.google.com/article/ALeqM5goQ9OvDd7cJ4EEDM2GWMRbyTFS-BwD8T7GMO80, accessed 25 December 2007.
34. World Business Council for Sustainable Development (2004).
35. Rappaport, A. and Creighton, S. (2007). *Degrees that matter, climate change and the university.* MIT Press.
36. *Ibid.*
37. Brown, M. (2009, March 8). Foes and costs halt coal plant expansions. *The Courier-Journal.* p. A4.
38. *Ibid.*
39. Cities go green (2008, September). *World's first smart grid.* p. 6.
40. *Ibid.*
41. *Ibid.*
42. Lewis, N. (2008). Powering the planet: Where in the world will our energy come from? *Energeia.* 10 (4). p. 1.
43. Huntsman, J. (2008, September 18). Presentation at the *Energizing Kentucky Conference*, Louisville, KY.
44. Lewis, N. (2008). Powering the planet: Where in the world will our energy come from? *Energeia.* 10 (4). p. 2.
45. Brown, M. (2009, March 8). Foes and costs halt coal plant expansions. *The Courier-Journal.* p. A4.
46. FOXNews.com (2009, April 17). EPA takes first step toward regulating pollution linked to climate change. http://www.foxnews.com/politics/first-100days/2009/04/17/epa-takes-step-regulating-pollution-linked-climate-change/, accessed 29 April 2009.
47. Ibid.
48. On April 2, 2007, the U.S. Supreme Court ruled in Massachusetts vs. EPA that carbon dioxide could be regulated as a pollutant under the Clean Air Act and that the government had a responsibility to issue a determination based on science. The G.W. Bush administration failed to act on this ruling.
49. FOXNews.com (2009, April 17). EPA takes first step toward regulating pollution linked to climate change. http://www.foxnews.com/politics/first-100days/2009/04/17/epa-takes-step-regulating-pollution-linked-climate-change/, accessed 29 April 2009.
50. Shin, L. (2009, February 18). EPA review reverberates through U.S. energy industry. http://solveclimate.com/blog/20090218/epa-review-reverberates-through-u-s-energy-industry/, accessed 29 April 2009.
51. Ibid.
52. Ibid.
53. Ibid.

Bibliography

Ambrose, W., et al. "Source-Sink Matching and Potential for Carbon Capture and Storage in the Gulf Coast." *The Gulf Coast Carbon Center.*

American Institute of Architects. "Sustainable Architectural Practice Position Statement." http://www.aia.org/SiteObjects/files/sustain_ps.pdf.

American Public Transportation Association. "Fact Sheet/Transit News." (March 2008). http://www.apta.com/.

American Society of Heating, Ventilations and Air-Conditioning Engineers. "90.1 Energy Standard for Buildings Except Low-Rise Residential Buildings."

American Wind Energy Association. "Installed U.S. Wind Power Capacity Surged 45% in 2007." (January 17, 2008). http://www.awea.org/newsroom/releases/AWEA_Market_Release_Q4_011708.html.

Andrews, R. L. *Managing the Environment, Managing Ourselves.* New Haven, Connecticut: Yale University Press, 1999.

Austin City Connection. "Wynn Announces Austin Climate Protection Plan." (February 7, 2007). http://www.ci.austin.tx.us/council/mw_acpp_release.html.

"Austin Climate Protection Plan." *The Austin Chronicle* (March 7, 2007). http://www.austinchronicle.com/gyrobase/Issue/story?oid=oid%3A453480.

Avery, G. *Leadership for Sustainable Futures.* Edward Elgar Publishing Limited, 2005.

Bailey, J. "Get the Energy Future Right." *The Courier-Journal* (January 5, 2009).

Bailey, L. "Public Transportation and Petroleum Savings in the U.S.: Reducing Dependence on Oil." ICF International (January 2007). http://www.apta.com/research/info/online/documents/apta_public_transportation_fuel_savings_final_010807.pdf.

Bank of America. "Bank of America Commits $20 Billion to Green Lending." (March 6, 2007). http://blogs.business2.com/greenwombat/2007/03/bank_america_co.html.

Barringer, F. "Urban Areas on West Coast Produce Least Emissions

Per Capita, Researchers Find." *New York Times* (May 29, 2008). http://www.nytimes.com/2008/05/29/us/29pollute.html?ref=us.

Basin Electric Power Cooperative. "Harper Weighs in on Capture and Storage." (September 17, 2007). http://www.basinelectric.com/NewsCenter/News/Briefs/Harper_weighs_in_on_html.

Berman, J. "Green Logistics: DPWN Sets Goal of 30 Percent Carbon Footprint Reduction Through 'GoGreen' Initiative." (April 16, 2008). http://www.scmr.com/article/CA6551866.html.

Bloomburg News. "Companies Join Effort to Curb Emissions." *USA Today* (January 21, 2008).

Bonts, M. "Florida's First Zero-Energy, LEED Platinum Certified Home Unveiled in Panama City." (March 26, 2008). http://www.usgbc.org/Docs/News/Stalwart%20LEED%20Release_AK.pdf.

Borenstein, S. "Report: Global Warming 'Very Likely' Man-Made, Unstoppable for Centuries." *The Courier-Journal* (February 6, 2007).

Boucher, M. "Resource Efficient Buildings – Balancing the Bottom Line." Proceedings of the 2004 World Energy Engineering Congress. Atlanta: Association of Energy Engineers, 2004.

Breslau, K. "The Green Giant." *Newsweek* (April 16, 2007).

Brown, M. "Foes and Costs Halt Coal Plant expansions." *The Courier-Journal* (March 8, 2009).

Brown, M. "Shrinking the Carbon Footprint of Metropolitan America." Brookings Institution Metropolitan Policy Program. (May, 2008).

Bruggers, J. "EPA Proposes First Carbon Storage Rules." *The Courier-Journal* (July 16, 2008).

Bruggers, J. "Our Changing Climate – Unprecedented Study Examines 60 Years of Weather." *The Courier-Journal* (October 20, 2008).

Bruggers, J. "Retrofit Hydro Plant Offers Clean Power." *The Courier-Journal* (October 28, 2008).

Bryant, S. "Geologic CO_2 Storage – Can the Oil and Gas Industry Help Save the Planet?" *Distinguished Author Series* (September, 2007).

California Air Resources Board. "AB32 Fact Sheet – California Global Warming Solutions Act of 2006." (September 25, 2006). http://www.arb.ca.gov/cc/factsheets/ab32factsheet.pdf.

California Public Services Commission. "PUC Creates Groundbreaking Solar Energy Program." (January 12, 2006). http://docs.cpuc.ca.gov/published/News_release/52745.htm.

California Public Utilities Commission. "CPUC Establishes Institute for Climate Solutions to Build on State's Environmental Leadership."

(April 10, 2008). http://docs.cpuc.ca.gov/Published/News_release/81168.htm.

Cambridge Community Development. "Why Take Local Action on Climate Change?" (2004-2007). www.cambridgema.gov/cdd/et/climate/index.html.

Canon. "Sustainability Is Our Standard For Measuring CO$_2$ Reduction." *Newsweek* (September 17, 2007).

Capehart, B. *Encyclopedia of Energy Engineering and Technology.* Boca Raton: CRC Press. 2007.

Carbon Positive. "What are CERs?" (September 12, 2005). http://www.carbonpositive.net/viewarticle.aspx?articleID=34.

Carbon Positive. "Kyoto's CDM Approaches the Crossroads." (September 13, 2006). http://www.carbonpositive.net/viewarticle.aspx?articleID=423.

Card, R. "Public-Private Partnership for Technology Innovation." Remarks by the Undersecretary of Energy (U.S.) at a roundtable discussion on Technology, Including Technology Use and Development and Transfer of Technologies. (December 11, 2003).

Carr, M. "EU, UN Emissions Systems to be Linked by November 17." *Bloomberg* (August 30, 2007). http://www.bloomberg.com/apps/news?pid=newsarchive&sid=aGZeMMQ1Gok0.

Cascadia Consulting Group, Inc. "Carbon Reduction Strategies: Choosing the Right Path for Your Business." (October, 2007).

Cashman, K. and Forem, J. *Awakening the Leader Within: A Story of Transformation.* Wiley & Sons, Inc., 2003.

CBC News. "Harper's Letter Dismisses Kyoto as 'Socialist Scheme'. (January 30, 2007). http://www.cbc.ca/canada/story/2007/01/30/harper-kyoto.html.

Center for Energy and Environmental Policy. *Carbon Dioxide Emission Reduction Technologies and Measures in U.S. Industrial Sector.* University of Delaware, 2007.

Chapa, J. "First LEED Platinum Carbon Neutral Building." (November 6, 2007). http://www.inhabitat.com/2007/11/08/first-leed-platinum-carbon-neutral-building/.

Chasek, P. *The Global Environment in the Twenty-First Century: Prospects for International Cooperation.* New York: The United Nations University, 2000.

Chevrolet. "Concept Chevy Volt." http://www.chevrolet.com/electric-car/.

Chourci, N. I. (ed.) *Global Accord*. Cambridge, Massachusetts: The Massachusetts Institute of Technology Press, 1993.

City of Austin. "Austin Climate Protection Plan." (2007). http://www.ci.austin.tx.us/council/downloads/mw_acpp_points.pdf.

City of Santa Monica. "Santa Monica Honored as Sustainability Leader in U.S." (June 2, 2005). http://www.smgov.net/news/releases/archive/2005/epwm20050602.htm.

City of Santa Monica. "Santa Monica Urban Runoff Recycling Facility." (2007). http://www.smgov.net/epwm/smurrf/smurrf.html.

City of Santa Monica. "Sustainable Santa Monica." (2008). http://www.smgov.net/epd/scp/index.htm.

Clayton, K. "The Energy Manager's Role in Sustainable Design." *Solutions for Energy Security and Facility Management Challenges*. Atlanta, Georgia: Association of Energy Engineers, 2003.

Clayton, M. "Algae – Like a Breath Mint for Smokestacks." *The Christian Science Monitor*. (January 11, 2006). http://www.csmonitor.com/2006/0111/p01s03-sten.htm.

Climate Change. "Conversion of 1 MMT CO_2 to Familiar Units." (September 25, 2006). http://arb.ca.gov/cc/factsheets/1mmtconversion.pdf.

Climate Change Projects Office, United Kingdom Department of Trade and Industry. "Carbon prices." (April, 2005).

"Climate Connections." Maps and documents. *National Geographic* (October, 2007).

Cohen, K. "Letters." *Newsweek* (September 3, 2007).

Colonbo, U. "Development and the Global Environment." In Hollander J. M. (ed.), *The Energy-Environment Connection*. Washington, D.C.: Island Press, 1992.

Commonwealth of Australia, Department of Climate Change. "Carbon Pollution Reduction Scheme Green Paper." (July, 2008).

Co-Op America. "Carbon Sequestration." (2005). http://www.coopamerica.org/programs/climate/dirtyenergy/coal/ccs.cfm.

Copeland, L. "State Workers in Utah Shifting to 4-day Week." *USA Today* (July 1, 2008).

Copeland, M. and McNichol, T. "Here Comes the Sun." *Business 2.0* (November, 2006).

"Corporate Leaders Back Action on Global Warming." *The Courier-Journal* (January 1, 2007).

Covey, S. *Principle-Centered Leadership*. Salt Lake City, Utah: Franklin

Covey Co., 2007.

CRed. "The Community Carbon Reduction Project at UNC – Chapel Hill." (2005). http://www.ie.unc.edu/content/research/cred/cambridge.html.

Cusimano, M. (ed.). *Beyond Sovereignty Issues for a Global Agenda*. Boston, Massachusetts: Bedford/ St. Martin's, 2000.

David Suzuki Foundation. "Carbon Tax Backgrounder – B.C. Budget 2008." (2008). http://www.davidsuzuki.org/files/climate/Briefing_Note_-_BC_Budget_2008.pdf.er 4, 2007).

Davidson, P. "Clean Energy Can't Meet Growing Demand." *USA Today* (October 4, 2007). http://www.usatoday.com/money/industries/energy/environment/2007-10-03-clean-energy_N.htm?csp=34.

Davidson, P. "Coal King Peabody Cleans Up." *USA Today* (August 19, 2008).

Davidson, P. "4 Creative Solutions to Energy Problems." *USA Today* (September 8, 2008).

Davidson, P. "Green Exchange Takes Root Soon." *USA Today* (December 13, 2007).

Davidson, P. "Greenhouse Gas Villain Rehabbed." *USA Today* (February 25, 2009).

Davidson, P. "New Battery Packs Power Punch." *USA Today* (July 5, 2007).

Davidson, P. "TXU Takes $32B Deal, Cuts Coal Plans." USA Today. (February 26, 2007). p.B1.

Department of State Publications. "Asia-Pacific Partnership on Clean Development and Climate." (2007).

Deutsche Post World Net. "Accepting Our Social Responsibilities." (2008). http://www.dpwn.de/dpwn?skin=hi&check=yes&lang=de_EN&xmlFile=300000032.

Deutsche Post World Net. "Deutsche Post World Net Starts Global Climate Protection Program GoGreen." (April 8, 2008). http://www.dpwn.de/dpwn?tab=1&skin=hi&check=yes&lang=de_EN&xmlFile=2009784.

Dilling, L. and Moser, S. (eds.). *Creating Climate for Change: Communicating Climate Change and Facilitating Social Change*. Cambridge: Cambridge University Press, 2007.

Doppelt, B. *Leading Change Toward Sustainability: A Change-Management Guide for Business, Government, and Civil Society*. Sheffield, South Yorkshire, UK: Greenleaf Publishing Limited, 2003.

Dowdeswell, E. "United Nations Environment Program." In speech celebrating World Environment Day (1996). http://ces.iisc.ernet.in/hpg/envis/worlenvdoc64.html.

Dreessen, T. "Scaling Up Energy Efficiency Financing." Paper presented at Wilton Park, UK. (July 18, 2008).

Duke Energy. "Eight Utilities Seek to Increase Energy Efficiency Investment $500 Million Annually." (September 27, 2007). http://www.duke-energy.com/news/releases/2007092701.asp.

Dunn, C. "Dutch Bank Introduces 'Climate Credit Card'." (October 2, 2006). http://www.treehugger.com/files/2006/10/dutch_bank_intr.php.

Durst, S. "Problem No. 9: Waste Disposal." Business 2.0. (January 26, 2007). http://money.cnn.com/2007/01/24/magazines/business2/Prob9_Wastedisposal.biz2/index.htm, accessed 2 May 2009.

Earth Policy Institute. "U.S. Mayors Pledge to Cut Greenhouse Gases While Bush Administration Takes No Action." (May 12, 2006). http://www.citymayors.com/environment/usmayors_kyoto.html.

Ecofys UK. "Carbon Offsetting, A Report for the Arts Council of England." (February 19, 2007). http://www.arts.org.uk/documents/projects/phpVg6Scb.pdf.

Edemariam A. "Dark Side of the Boom." *Postmagazine*, Hong Kong (December 9, 2007): 28-32.

Eilperin, J. "Faster Climate Change." *The Courier-Journal* (January 5, 2009).

Eliason, D. and Perry, M. "CO_2 Recovery and Sequestration at Dakota Gasification Company." (October, 2004). http://www.gasification.org/Docs/Conferences/2004/11ELIA_Paper.pdf.

Elliott, L. *The Global Politics of the Environment.* New York: New York University Press, 1998.

Energy Information Administration. "Annual Energy Outlook 2004 with Projections to 2025." DOE/EIA -0554. (February, 2004).

Energy Information Administration. "Emissions of Greenhouse Gases in the United States, Executive Summary." (2005). www.eia.doe.gov/oiaf/1605/ggrpt/pdf/executive_summary.pdf.

Energy Information Administration. "Greenhouse Gases, Climate Change and Energy." http://www.eia.doe.gov/oiaf/1605/ggccebro/chapter1.html.

Energy Information Administration. "India: Environmental issues (2004)" and "China Environmental Issues (2003)," Country

Analysis Briefings. (2003-2004).

Energy Information Administration. "International Energy Annual 2005." (September 18, 2007). http://www.eia.doe.gov/pub/international/iealf/tableh1co2.xls.

Energy Information Administration. "International Energy Outlook 2007." (May, 2007). http://www.eia.doe.gov/oiaf/ieo/.

Energy Solutions Operation. "Public Transportation's Contribution to U.S. Greenhouse Gas Reduction." Science Applications International Corporation. (September, 2007).

Energy Vortex. "CPUC Releases Updated Energy Action Plan Focused on Climate Change." http://www.energyvortex.com/pages/headlinedetails.cfm?id=3416.

Engeler, E. "Antarctic Glaciers Melting Faster Than Thought, Report Says." *The Courier-Journal* (February 26, 2009).

Environment News Service. "Buildings of the Future Energy Self-Sufficient, Carbon Neutral." (March 29, 2006). http://www.ens-newswire.com/ens/mar2006/2006-03-29-03.asp.

Environmental Leader. "Toyota Releases Environmental Report, Plans 2.2 MW Solar Power System." (December 7, 2007). http://www.environmentalleader.com/2007/12/07/toyota-releases-environmental-report-plans-22-mw-solar-power-system/.

EurActiv. "EU Emissions Trading Scheme." (June 29, 2007). http://www.euractiv.com/en/sustainability/eu-emissions-trading-scheme/article-133629.

Ford, W. (2002). Speech by the Chairman and CEO of Ford Motor Company.

Fowler, J. "U.N. Agency Says 2003 Third-Hottest on Record." *The Courier-Journal* (December 17, 2003).

FOXNews.com. "EPA Takes First Step Toward Regulating Pollution Linked to Climate Change." (April 17, 2009). http://www.foxnews.com/politics/first100days/2009/04/17/epa-takes-step-regulating-pollution-linked-climate-change/.

Friedman T. *Hot, Flat and Crowded*. New York: Farrar, Straus and Giroux, 2008.

FutureGen Alliance. "FutureGen Alliance Hails Senate Appropriations Committee for Protecting FutureGen at Mattoon Funding." (July 11, 2008). http://www.futuregenalliance.org/news/releases/pr_07-11-08.pdf.

Gale, S. "Carbon Reduction Wins Mega Brand Attention at Conference." *GreenBiz* (September 14, 2007). http://www.greenbiz.com/news/news_third.cfm?NewsID=35906.

Geri. "U.S. Mayors Embrace Kyoto Protocol." *Good News Network* (October 23, 2006). http://www.goodnewsnetwork.org/earth/general/cities-embrace-kyoto.html.

Global Energy Technology Strategy Program. "Research Programs." (2005). http://www.pnl.gov/gtsp/research/model.stm.

Global Environment Facility. "About the GEF." (2007). http://www.gefweb.org/interior.aspx?id=50.

Goddard Space Flight Center. "Climate Change May Become Major Player in Ozone Loss." (June 4, 2002). www.gsfc.nasa,gov/topstory/2002041greengas.html.

Gore, A. *The Assault on Reason.* New York: Penguin Press, 2007.

Gore, A. "Making Peace with the Planet." Nobel Peace Prize. (2007).

Governor's Green Government Council. "What Is a Green Building?" *Building Green in Pennsylvania.* http://www.gggc.state.pa.us/gggc/lib/gggc/documents/whatis041202.pdf.

Governor's Office of Energy Policy. "Possible and Profitable: Energy Efficiency Investments in the Building Sector." *Kentucky Energy Watch* (April 17, 2007): 8 (16).

Governor of the State of California. "Executive Order S-3-05." Signed by Arnold Schwarzenegger. (June 1, 2005). http://gov.ca.gov/executive-order/1861/.

Graham-Harrison, E. "World Carbon Market Leaps in 2006, China Share Down." (October 17, 2006). http://www.planetark.org/dailynewsstory.cfm/newsid/38696/story.htm.

Grumman, D. L. (ed.). *ASHRAE Green Guide.* Atlanta: ASHRAE, 2003.

Hackman, R. "Leading Teams." (2004).

Hakim, D. "Ford Lays Out a Move to Cut Auto Emissions." (October 2, 2004). http://www.pinnaclecng.com/pdfs/Ford%20Cut%20Emissions.pdf.

Hanley, C. "Malaria's Spread on Island Blamed on Global Warming." *The Courier-Journal* (December 9, 2007).

Hanley, C. "U.S. 'Not Ready' to Commit to Emission Cuts at Meeting." *The Courier-Journal* (December 9, 2007).

Hansen, J. "NASA's Goddard Institute for Space Studies." New York. (2006).

Hansen. S. *Performance Contracting: Expanding Horizons.* Lilburn, Georgia: The Fairmont Press, 2006.

Harris, E. "Felling of Rain Forests Imperils Climate." (February 3, 2008). http://www.commercialappeal.com/news/2008/feb/03/felling-of-rain-forests-imperils-global-climate/.

Harvey, F. and Fidler, S. "Industry Caught in 'Carbon Smokescreen'." *Financial Times* (April 25, 2007). http://www.ft.com/cms/s/0/48e334ce-f355-11db-9845-000b5df10621.html.

Hawksworth, J. *The World in 2050: Implications of Global Growth for Carbon Emissions and Climate Change Policy.* Pricewaterhouse-Coopers, 2006.

Hayes, D. "Getting Credit for Going Green." (February 28, 2008). http://www.americanprogress.org/issues/2008/03/carbon_offsets_report.html.

Healey, J. "Judge Says States Can Regulate Emissions." *USA Today* (September 13, 2007).

Healey, J., Woodyard, C. and Carty, S. "Alternative Power Sources for Autos Drive Into Spotlight." *USA Today* (September 1, 2007): 3B.

Hebert, H. "Can We Fix Global Warming?" *The Courier-Journal* (June 2, 2008).

Herbert, S. "The Spectrum of High Performance Buildings." *Cities Go Green* (September, 2008).

Hogan, D. "A Sea of Change." *The Courier-Journal* (December 8, 2008).

Holahan, D. "Just Warming Up." *Connecticut Magazine* (November, 2007).

Holland, A., Lee, K., and McNeil, D. *Global Sustainable Development in the 21st Century.* Edinburgh: Edinburgh University Press, 2000.

Hotz, R. "Wanted: Carbon Emissions." *The Courier-Journal* (February 11, 2007).

Hough, M. Cities and Natural Process. New York: Routledge, 1995.

Howard, B. "Net Positive Energy Homes: Maximizing Performance Using Advanced Software Systems." Conference paper presented at AEE Globalcon, Austin, Texas. (March, 2008).

HSBC. "Reducing Emissions." (2007). http://www.hsbccommittochange.com/environment/hsbc-case-studies/carbon-dioxide/reducing-emissions/index.aspx.

Hultman, N. "Carbon Sequestration." In the *Encyclopedia of Energy Engineering and Technology.* Editor – Barney Capehart, Ph.D. Lilburn, GA: The Fairmont Press, 2007.

Huntsman, J. In a keynote address at the *Energizing Kentucky Conference,* Louisville, Kentucky. (September 18, 2008).

iiSBE. "International Initiative for a Sustainable Built Environment."
    (2005). http://iisbe.org/iisbe/start/iisbe.htm.
Intergovernmental Panel on Climate Change. *Carbon Dioxide Capture
    and Storage.* Cambridge: Cambridge University Press, 2005.
Intergovernmental Panel on Climate Change. "Fourth Assessment Re-
    port, Mitigation of Climate Change." (November, 2007). http://
    www.ipcc.ch/ipccreports/ar4-wg3.htm.
Intergovernmental Panel on Climate Change. "Second Assessment
    Report." (1995). http://www.ipcc.ch/pdf/climate-changes-1995/
    ipcc-2nd-assessment/2nd-assessment-en.pdf.
International Design Center for the Environment. "eLCie - Building a
    Sustainable World." www.elcie.org.
International Emissions Trading Association. "IETA Position Paper on
    EU ETS Market functioning." (October 13, 2006). http://www.
    ieta.org/ieta/www/pages/getfile.php?docID=1926.
International Iron and Steel Institute. "Steel Industry Launches Global
    CO_2 Emissions Data Collection Programme." (April 15, 2008).
    http://www.worldsteel.org/index.php?action=newsdetail&id=238.
International Soil Tillage Research Organization (ISTRO). "Reversing
    Emissions From Land Use Change," Chapter 25. Found in *The
    Stern Review.* (2007)
Jhaveri, A. *Effective Leadership for Sustainable Development in the Public
    Sector.* Doctoral Dissertation. 2006.
John G. "Nuclear Energy Renaissance Worries Some Observers." *The
    Courier-Journal* (January 13, 2008).
Johnson, E. "Arlington County, Virginia." *Cities Go Green* (September,
    2008).
Johnson, L. "Building Design Leaders Collaborating on Carbon Neu-
    tral Buildings by 2030." (May 7, 2007). http://blog.fastcompany.
    com/archives/2007/05/07/building_design_leaders_collaborat-
    ing_on_carbonneutral_buildings_by_2030.html.
Kaempf, D. "U.S. Department of Energy's Industrial Technologies
    Program." (2007).
Kaihla, P. "Global Warming." Business 2.0 (January, 2007) 8 (1).
Kaltschnitt, M., Streicher, W. and Wiese, A. *Renewable Energy: Technol-
    ogy, Economics and Environment.* Springer-Verlag, 2007.
Kansas Corporation Commission. "Kansas Wind Resources Map."
    (April 11, 2008).
Karanowski, S. "Algae Fueling New Renewable Energy Research." *The

Courier-Journal (December 9, 2007).

Karanowski, S. "There's Oil in That Slime." *Associated Press* (November 29, 2007). http://ap.google.com/article/ALeqM5goQ9OvD-d7cJ4EEDM2GWMRbyTFSBwD8T7GMO80.

Kaufman, M. "Global Warming and Hot Air." (February 7, 2007).

Kaufman, M. "Scientists Rate Costs of Reducing Global Warming." *The Courier-Journal* (May 5, 2007).

KCPW. "Governor's Mansion, SUV Go Green." (2006). http://www.kcpw.org/article/3674.

Kennedy, J. "Nation's Space Effort." Address at Rice University, Houston, Texas (September 12, 1962). For transcript see "We Choose to Go to the Moon." http://www.learnetix.de/home/de/htdocs/Englisch/Moonpalace/Pdf/General/TheSpaceProgram.pdf.

Kerry, J. and Kerry, T. *This Moment on Earth: Today's New Environmentalists and Their Vision for the Future.* New York: Public Affairs, 2007.

"Kilowatt Killers." *Public Works Magazine* (February, 2009). http://www.pwmag.com/industry-news.asp?sectionID=0&articleID=879934.

Konch, W. "Builder Codes Turn a Green Leaf." *USA Today* (August 7, 2008).

Koomey, J., Weber, C., Atkinson, C., and Nicholls, A. "Addressing Energy-Related Challenges for the U.S. Buildings Sector: Results From the Clean Energy Futures Study." *Energy Policy.* 29 (14) (2001):1209-1221. http://www.autobloggreen.com/2007/08/21/ups-has-goals-for-reducing-their-fuel-use-and-emissions-how-ar/.

Korzeniewski, J. "UPS Has Goals for Reducing Fuel Use and Emissions." (August 21, 2007). http://www.autobloggreen.com/2007/08/21/ups-has-goals-for-reducing-their-fuel-use-and-emissions-how-ar/.

Kraft, M. *Environmental Policy and Politics*, Third Edition. New York: Pearson/Longman, 2004.

Kuck, S. "The Geography of America's Carbon Footprint." (May 29, 2008). http://www.worldchanging.com/archives/008068.html.

Lawrence Berkeley Laboratory. "Ocean fertilization." http://www-esd.lbl.gov/CLIMATE/OCEAN/fertilization.html.

Leitmann, J. Sustaining Cities: Environmental Planning and Management in Urban Design. McGraw-Hill. 1999.

Lewis, N. "Powering the Planet: Where in the World Will Our Energy Come From?" *Energeia.* (2008): 10 (4).

Lovejoy, T. "The Ocean's Food Chain Is At Risk." *Newsweek* (April 16, 2007).

Lubin, J. "Environment Attracts Corporate Interest." *The Courier-Journal* (August 17, 2008).

Lynch, D. "China's 'Grappling With One Hell-Of-A Problem.'" *USA Today* (September 18, 2007).

Lynch, D. "Opportunity Shines in Hazy Days of China." *USA Today* (September 18, 2007).

Lynn, P., Murtishaw, S. and Worrell, E. "Evaluation of Metrics and Baselines for Tracking Greenhouse Gas Emissions Trends: Recommendations for the California Climate Action Registry." *Lawrence Berkeley National Laboratory* (June, 2003).

Mansfield. D. "TVA Sees Future in Nuclear Power." *The Courier-Journal* (May 5, 2007).

Markels, M. and Barber, R. "Sequestration of CO_2 by Ocean Fertilization." Presentation for NETL Conference on Carbon Sequestration (May 14, 2001).

Martinez, F. Presentation at the NASEO Annual Conference. Overland Park, Kansas. (September 8, 2008).

Masdar. "Masdar Headquarters to be Located In World's First 'Positive Energy' Mixed-Use Building." (February 22, 2008). http://www.prnewswire.com/cgibin/stories.pl?ACCT=104&STORY=/www/story/02-22-2008/0004760606&EDATE.

Matis, C. "Discussion of Recent Changes to ASHRAE 90.1." NASEO Conference. Overland Park, Kansas. (September 8, 2008).

Mayor's Office of Communications. San Francisco's "Mayor Newsom Unveils First-Ever Carbon Offsets to Fight Global Warming." (December 18, 2007). www.sfgov.org/site/mayor_index.asp?id=72509.

McKibben, B. "Carbon's New Math." *National Geographic* (October, 2007). http://www.mckinsey.com/clientservice/ccsi/greenhouse-gas.asp.

McKinsey & Company. "Curbing Global Energy Demand Growth: The Energy Productivity Opportunity." (May, 2007). http://www.mckinsey.com/mgi/publications/Curbing_Global_Energy/index.as.

McKinsey & Company. "Reducing U.S. Greenhouse Gas Emissions: How Much at What Cost?" U.S. Greenhouse Gas Abatement Mapping Initiative. (December, 2007). http://www.mckinsey.

com/clientservice/ccsi/greenhousegas.asp.

Mega, V. *Sustainable Development, Energy and the City: A Civilization of Visions and Actions.* New York: Springer, 2005.

Metz, B. *Climate Change 2007.* Cambridge: Cambridge University Press, 2007.

Miller, P. "Saving Energy It Starts at Home." *National Geographic.* (March, 2009).

MGA Energy Summit. "Energy Security and Climate Stewardship Platform for the Midwest." (November 15, 2007).

Moser, S. and Dilling, L. Creating a Climate for Change. Cambridge: Cambridge University Press, 2007, p266

Mufson. S. "Federal Loans Fuel New Coal Plants." *The Courier-Journal* (May 14, 2007).

Muller, P. "Syngas in Johnson County." http://mypc.press-citizen.com/story.php?id_stories=304.

Muraya, N. "Austin Climate Protection Plan - Possibly the Most Aggressive City Greenhouse-Gas Reduction Plan." *AEE - Energy Engineering.* (2008): 105 (2).

Muraya, N. "Austin Climate Protection Plan." (November, 2008).

National Energy Technology Laboratory. "What Is Carbon Sequestration?" http://www.netl.doe.gov/technologies/carbon_seq/FAQs/carbon-seq.html.

National Oceanic and Atmospheric Association. "NOAA/ERSL Releases Ozone Depleting Gas Index." (January 15, 2007). www.noaa.gov/hotitems/storyDetail_org.php?sid=3843.

Naughton, K. "The Man Who Revived the Electric Car." *Newsweek* (December 31, 2007).

New York State Public Service Commission. "Historic Energy Efficiency Program Gets Underway in NY." (July 19, 2008). Tdworld.com/customer_service/ny-energy-efficiency-0806.

Newman, P. and Jennings, I. Cities as Sustainable Ecosystems. Washington, D.C.: Island Press, 2008.

Nunez, A. "EPA Unveils Hydraulic Hybrid UPS Delivery Truck." (June 26, 2006). http://www.autoblog.com/2006/06/26/epa-unveils-hydraulic-hybrid-ups-delivery-truck/.

O'Dell, J. "Honda Unveils 'Super-Clean Diesel Engine'." *The Los Angeles Times* (September 25, 2006).

Office of the Presidential Press Secretary. "Executive Order: Strengthening Federal Environmental, Energy, and Transportation Management." (January 24, 2007). http://www.whitehouse.gov/news/releases/2007/01/print/20070124-2.html.

Peirce, N. "A 'Green' Rx to Save Carbon: City Density Plus Transit." *The Courier-Journal* (May 6, 2008).

Pew Center on Global Climate Change. "Climate Change 101 – Local Action." http://www.pewclimate.org/docUploads/Climate-101-LocalBlueline.pdf.

Pew Center on Global Climate Change. "Impact of the Climate Change Program on Industrial CO_2 Emissions." (August, 2003).

Pew Center on Global Climate Change. "Regional Initiatives." (July 10, 2008).

Pew Center on Global Climate Change. "Summary of the Lieberman-McCain Climate Stewardship Act." (2003). http://www.pewclimate.org/policy_center/analyses/s_139_summary.cfm.

Philips, M. "The Fading Forests of the Sea." *Newsweek* (July 9, 2007).

Pierantozz1, R. "Carbon Dioxide." *Kirk-Othmer Encyclopedia of Chemical Technology.* Wiley, 2001.

Porretto, J. and Wardell, J. "Agency Warns of Energy Crisis." *The Courier-Journal* (November 13, 2008).

Portland on Line. "Global Warming Progress Report." (June, 2005). www.portlandonline.com.

Portney, K.E. Taking Sustainable Cities Seriously. Cambridge, Massachusetts: The MIT Press, 2003.

"Power Execs: Cap Carbon Emissions." (October 22, 2006). http://www.wired.com/techbiz/media/news/2006/10/71990.

Rabinowitz, G. "Modern Model T Arrives." *The Courier-Journal* (January 11, 2008): D1.

Rainwater, B. "Local Leaders in Sustainability – A Study of Green Building Programs in Our Nation's Communities." (2007). http://www.aia.org/SiteObjects/files/LLinSustain(Findings)_Final.pdf.

Rappaport, A. and Creighton, S. *Degrees That Matter, Climate Change and the University.* Boston: MIT Press, 2007.

Renewable Energy Access. "700 Mw From Electricity to Come From Landfill Gas." (June 27, 2007). http://www.renewableenergyaccess.com/rea/news/story?id=49123.

Revkin, A. "Poor Nations Bear Brunt of Warming." *New York Times* (April 1, 2007).

Rhode, R. "Global Warming Art." http://www.globalwarmingart.com/

wiki/Image:1000_Year_Temperature_Comparison_png.

Ritter, J. "California Sees Sprawl As Warming Culprit." *USA Today* (June 6, 2007).

Robinson, S. "The Global Warming Survival Guide – 38. Trade Carbon for Capital." *Time* (April 14, 2007). www.time.com/specials/2007/environment/article/0,28804,1602354_163074_1603643,00.html.

Roland, E. and Molina, M. "The CFC Ozone Puzzle: Environmental Science in the Global Arena." Presented at the First National Conference on Science, Policy and the Environment. Washington D.C. (2001).

Roosa, S. *The Sustainable Development Handbook.* Lilburn GA: The Fairmont Press, 2008.

Roseland, M. Toward Sustainable Communities. Gabriola Island, B.C.: New Society Publishers, 1998.

Samuelson. P. "Greenhouse Guessing." (November 10, 2006). www.washingtonpost.com/wp-dyn/content/article/2006/11/09/AR2006110901768.html?nav=rss_opinion/columns.

Schloerer, J. "Why Does Atmospheric CO_2 Rise?" (October, 1996). http://www.radix.net/~bobg/faqs/scq.CO2rise.html.

Sedensky, M. "Mayors Pick Up Where Washington Failed on Kyoto." (October 25, 2006). http://www.iht.com/articles/2006/10/23/business/kyoto.php.

Shaffer, M. "Algae Could Be Fuel of the Future." *The Arizona Republic* (October 14, 2006). http://www.azcentral.com/arizonarepublic/business/articles/1014biz-algae1014.html.

Shin, L. "EPA Review Reverberates Through U.S. Energy Industry." (February 18, 2009). http://solveclimate.com/blog/20090218/epa-review-reverberates-through-u-s-energy-industry/.

Shulman, R. "Carbon Sale Raises $40 Million." *The Washington Post* (September 30, 2007).

Siemens AG. "Combat Climate Change – Less Is More." (2007). http://wap.siemens-mobile.com/en/journal/story2.html.

Skoloff, B. "Cities Offer Cash, Perks to 'Go Green'." *The Courier-Journal* (December 28, 2007).

Sommer, S. "Green Banking." *Gonzo Banker* (October 5, 2007). http://www.gonzobanker.com/article.aspx?Article=350.

Stangis, D. "Director of Corporate Responsibility for Intel Corporation." (September 14, 2007).

State of California. "Legislative Assembly Bill 32." (2006).

Stern Review. *The Economics of Climate Change*. Cambridge, MA: Cambridge University Press, 2006.

Takahashi, T. "Toyota Demonstrates Long-Distance Fuel-Cell Car." *The Courier-Journal* (September 29, 2007).

The Christensen Corporation. "Banner Bank Building." http://www. hdrinc.com/13/38/1/default.aspx?projectID=406.

The Climate Group. "Public-Private Partnerships: Local Initiatives 2007." (2007).

The Nature Conservancy. "First Conservation Initiative Certified for Reducing Greenhouse Gas Emissions." (December, 2005). http:// www.nature.org./initiatives/climatechange/press/press2192.html.

The Netherlands Organization for Scientific Research. "Prepare CO_2 Capture and Storage Now for Greater Environmental Benefit Later." (April 10, 2007). http://www.nwo.nl/nwohome.nsf/pages/NWOA_6ZKL7V_Eng.

Tietenberg, T. *Environmental and Natural Resource Economics*. New York: Harper-Collins Publishers, 2000.

Tool Base Services. "The Zero Energy Homes Project." http://www. toolbase.org/ToolbaseResources/level4CaseStudies.aspx?ContentD etailID=2469&BucketID=2&CategoryID=58.

Trade and Environment Database. "Oil Production and Environmental Damage." http://www.american.edu/TED/projects/tedcross/xoilpr15.htm#r3, American University.

Triplepundit. "UPS Launches a Small Zero Emissions Fleet." (January 15, 2008). http://www.triplepundit.com/pages/ups-launches-a-small-zero-emis-002848.php.

UNESCO. "Climate Change Threatens UNESCO World Heritage Sites." (April 10, 2007). whe.unceso.org/pg_friendly_print. cfm?id=319&cid=82&.

Union of Concerned Scientists. "Clean Energy – Public Benefits of Renewable Energy Use." (1999).

United Nations Conference on Environment and Development. "The Earth Summit." (June, 1992). http://www.un.org/geninfo/bp/enviro.html.

United Nations Framework Convention on Climate Change. "Education and Outreach." (January 24, 2008). http://unfccc.int/cooperation_and_support/education_and_outreach/items/2529.php.

U.S. Conference of Mayors. "Climate Protection Agreement." (2005). http://www.usmayors.org/climateprotection/documents/

mcpAgreement.pdf.

U.S. Department of Energy. "Carbon Dioxide Emissions for U.S. buildings." *Buildings Energy Data Book.* 2007. http://www.btscoredatabook.net/?id=view_book&c=3.

U.S. Department of Energy. "Carbon Sequestration Focus Areas." http://cdiac2.esd.ornl.gov/scienceman.html#enchancing.

U.S. Department of Energy DER/CHP. "Federal Energy Management Program." (2006). http://www1.eere.energy.gov/femp/.

U.S. Department of Energy DER/CHP. *Greening Federal Facilities: An Energy, Environmental and Economic Resource Drive for Federal Facility Managers and Designers* (2nd ed.), 2001.

U.S. Department of Energy. "Industrial Technology Program." (2006). http://industrial-energy.lbl.gov/node/228.

U.S. Department of Energy. *2007 Buildings Energy Data Book.* (September, 2007). http://www.btscoredatabook.net/docs/2007-bedb-0921.pdf.

U.S. Environmental Protection Agency. "Be a Leader – Change Our Environment for the Better." Portfolio Manager Overview. www.energystar.gov/index.cfm?c=spp_res.pt_neprs_learn.

U.S. Environmental Protection Agency. "Inventory of U.S. Greenhouse Gas Emissions and Sinks/1990-2006." Public Review Draft. (February, 2008).

U.S. Environmental Protection Agency. "Measuring Energy Star Program Results." (November, 2005).

U.S. Environmental Protection Agency. "Mountaintop Mining / Valley Fills in Appalachia: Final Programmatic Environmental Impact Statement." (October, 2005).

U.S. Environmental Protection Agency. "Portfolio Manager Overview." www.energystar.gov/index.cfm?c=evaluate_performance.bus_portfolimanager.

U.S. Environmental Protection Agency. "Superior Energy Management Creates Environmental Leaders." *Guidelines for Energy Management, Overview.* http://www.energystar.gov/index.cfm?c=guidelines.guidelines_index.

U.S. Geological Survey. "Volcanic Gases and Their Effects." (January 10. 2006). http://volcanoes.usgs.gov/Hazards/What/VolGas/volgas.html.

U.S. Green Building Council. "LEED – Certification Process." (October, 2004). www.usgbc.org/LEED/Project/certprocess.asp.

U.S. Green Building Council. "Green Building Facts." (October, 2006). www.usgbc.org/DisplayPage.aspx?CategoryID= 2027.

URS Europe. http://www.urseurope.com/services/engineering/engineering-breean.htm. (February, 2005).

Vergano, D. and O'Driscoll, P. "Is Earth Nearing Its 'Tipping Points'?" *USA Today* (April 4, 2007).

Vestas. "Number 1 in Modern Energy." http://www.vestas.com/en/wind-power-solutions/wind-turbines/3.0-mw.

Vigar, D. "Climate Change: The Role of Global Companies." London: Tomorrow's Company, 2006.

Warner J. "Selling Carbon LFG Credits." *MSW Management* (November, 2007). http://www.mswmanagement.com/mw_0711_selling.html.

Watson, J. "U.N. Report: Climate Change to Have Global Impact." *The Courier-Journal* (April 11, 2007).

Wheeler, S. and Beatley, T. *The Sustainable Urban Development Reader.* New York: Routledge, 2004.

Wikipedia. "Methane." http://en.wikipedia.org/wiki/Methane.

Wikipedia. "Ozone layer." http://en.wikipedia.org/wiki/Ozone_layer.

Wikipedia. "Plasma Arc Waste Disposal." http://en.wikipedia.org/wiki/Plasma_arc_gasification.

Wolfe, B. "UPS Converting to Biodiesel for Pilot Project." *The Courier-Journal* (October 9, 2007).

Worrell, E., Martin, N. and Price. L. "Energy Efficiency and Carbon Emissions Reduction Opportunities in the U.S. Iron and Steel Industry." *Industrial Energy Analysis.* (1999). http://industrial-energy.lbl.gov/node/363.

Yeboah, F. Yegulalp, T. and Singh, H. "Future Zero Emissions Carbon Technology – Valuation and Policy Issues." *AEE - Energy Engineering.* (2007): 105 (2).

Young, D. "Building on What We've Built." *Preservation* (January, 2008).

Yu, R. Airports Go Green with Eco-Friendly Efforts. *USA Today* (September 17, 2008).

Zaks, D. "Geoplasma–Plasma Arc Incineration." (September 12, 2006). http://www.worldchanging.com/archives/004926.html.

Zhang, Z. "The Challenging Economic and Social Issues of Climate Change: Introduction." *Energy Policy* (2004).

Zinga, S. Presentation at the Energizing Kentucky Conference, Louisville, Kentucky. (September 18-19, 2008).

Appendix

ABBREVIATIONS

3Es	Energy, Environment, Economy
AAU	Assigned Amount Unit
ACPP	Austin Climate Protection Plan
AEE	Association of Energy Engineers
AIA	American Institute of Architects
AIRE	Arlington Initiative to Reduce Emissions
APP	Asia-Pacific Partnership
APS	Arizona Public Service
APTA	American Public Transportation Association
ARB	Air Resources Board
ARC	Assessment Recommendation Code
ASHRAE	American Society of Heating, Refrigerating & Air-Conditioning Engineers
BAU	Business-As-Usual
BedZED	Beddington Zero Energy Development
BREEAM	Building Research Establishment Environmental Assessment Method
Btu	British Thermal Units, equal to 1.055 kilojoules
C	Celsius
CCS	Carbon Capture and Storage
CaCO$_3$	Calcium Carbonate
CAR	Climate Action Report
CCA	Climate Change Agreement
CCL	Climate Change Levy
CCX	Chicago Climate Exchange
CDC	Carbon Disclosure Project
CDM	Clean Development Mechanism
CEO	Chief Executive Officer
CER	Certified Emissions Reductions
CERUPT	Certified Emissions Reduction Unit Procurement Tender

CFC   Chlorofluorocarbons
CFI   Carbon Financial Instrument
CFL   Compact Fluorescent Lamps
CGE   Computational General Equilibrium
CH_4 Methane
CHP   Combined Heat and Power
CICS   California Institute for Climate Solutions
cm   centimeter
CO_2 e   Carbon Dioxide Emissions
CO_2 Carbon Dioxide
Co-ops   Cooperatives
CPUC   California Public Utilities Commission
CRed   Community Carbon Reduction Program
CSR   Corporate Social Responsibility

DER   Distributed Energy Resources
DHL   Dalsey, Hillblom, Lynn (Deutsche Post World Net Co., Germany)
DPWN   Deutsche Post Work Net
DSM   Demand-side Management

ECX   European Climate Exchange
EIA   Energy Information Administration
EMS   Energy Monitoring and Control System
ENUSIM   Energy End Use Simulation Model
EOL   Enhanced Oil Recovery
ERU   Emissions Reduction Unit
ESCO   Energy Service Companies
ETS   Emissions Trading Scheme
EU   European Union

F   Fahrenheit
FEMP   Federal Energy Management Program
ft   foot
Ft^2 Square foot

GDP   Gross Domestic Product
GE   General Electric
GEF   Global Environmental Facility

GESP	Global Energy Strategy Program
GHG	Greenhouse Gas
GW	Gigawatts

HFC	Hydrofluorocarbons
HVAC	Heating, Ventilating and Air Conditioning

IAEA	International Atomic Energy Agency
ICFI	ICF International, a Global Professional Services Firm
ICLEI	International Council for Local Environmental Initiative
IEA	International Energy Agency
IECC	International Energy Conservation Code
IET	International Emissions Trading
IETA	International Emissions Trading Association
IGCC	Integrated Gasification Combined Cycle
IISBE	International Initiative for a Sustainable Built Environment
ILA	International Leadership Association
IPCC	Intergovernmental Panel on Climate Change
IPMVP	International Performance Measurement and Verification Protocol
IRR	Internal Rate of Return
ISTRO	International Soil Tillage Research Organization
ITP	Industrial Technologies Program

JI	Joint Implementation

kBtu	Thousand British Thermal Units
Kcal	Kilocalorie
Kg	Kilogram
Kj	Kilojoule, equal to .9478 BTUs
Km	Kilometer
Km/l	Kilometers per liter
Km²	Square kilometer
Kph	Kilometers per hour
kW	Kilowatt
kW²	Kilowatt squared
kWh	Kilowatt hour

LBNL	Lawrence Berkeley National Laboratory
lbs	pounds
LCSWMA	Lancaster County Solid Waste Management Authority
LED	Light-Emitting Diodes
LEED	Leadership in Energy and Environmental Design
LFG	Landfill Gas
LNG	Liquefied Natural Gas
M	Meter
MBTA	Massachusetts Bay Transportation Authority
MDME[3]	Multi-Sector Dynamic Model Energy-Environment-Economy
MMT	Million Metric Tons
MMTCE	Million Metric Tons of Carbon Equivalent
Mpg	Miles per gallon
Mph	Miles per hour
MSW	Municipal Solid Waste
MW	Megawatt
N_2O	Nitrogen Oxide
NASA	National Aeronautics and Space Administration
NEMS	National Energy Modeling Systems
NGO	Non-Governmental Organization
NPE	Net Positive Energy
PART	Program Assessment Rating Tool
PCF	Prototype Carbon Fund
PFC	Perfluorocompounds
ppb	parts per billion
ppm	parts per million
PV	Photovoltaic
PVC	Polyvinylchloride
PWC	Pricewaterhouse-Coopers
QBTUs	1 Quadrillion Btus = 10 million billion Btus
R&D	Research and Development
RD&D	Research, Development and Demonstration
RFP	Requests for Proposals

RGGI Regional Greenhouse Gas Initiative
RMU Removal Unit
RPS Renewable Portfolio Standards
RWE Rheinisch – Westfalisches Electrizitatswerk (AG), a German Utility

SAIC Science Application International
SCC Social Costs of Carbon
SIC Standard Industrial Code
SMURF Santa Monica Urban Runoff Recycling Facility
SPC Supercritical Pulverized Coal
SUV Sport Utility Vehicle

T&D Transmission and Distribution
TAR Third Assessment Report
TOD Transit-Oriented Development
TPES Total Primary Energy Supply
TVA Tennessee Valley Authority

U.S. United States
UK United Kingdom
UN United Nations
UNDP United Nations Development Program
UPS Uninterruptible Power Supply
USDOE United States Department of Energy
USEPA United States Environmental Protection Agency
USGBC United States Green Building Council

VMT Vehicle miles traveled

WCI Western Climate Initiative
ZEB Zero Energy Buildings
ZEC Zero Emissions Coal
ZEH Zero Energy Homes

Index